Failure Analysis of Wood and Wood-Based Products

Failure Analysis of Wood and Wood-Based Products

Dirk Lukowsky

Translated by David Wade

McGraw Hill Education

New York Chicago San Francisco
Athens London Madrid
Mexico City Milan New Delhi
Singapore Sydney Toronto

Failure Analysis of Wood and Wood-Based Products

1 2 3 4 5 6 7 8 9 0 CTP/CTP 1 2 0 9 8 7 6 5 4

ISBN 978-0-07-183937-2
MHID 0-07-183937-2

This book is based on *Schadensanalyse Holz und Holzwerkstoffe*, published by Fraunhofer IRB Verlag, Stuttgart, Germany, copyright © 2013.

The pages within this book were printed on acid-free paper.

Sponsoring Editor
Robert Argentieri

Editorial Supervisor
Stephen M. Smith

Production Supervisor
Lynn M. Messina

Project Manager
Dheeraj Chahal, MPS Limited

Copy Editor
Rashi Sinha, MPS Limited

Proofreader
Gina Delaney

Art Director, Cover
Jeff Weeks

Composition
MPS Limited

About the Author

Dirk Lukowsky, Ph.D., started his career as a finish carpenter, eventually qualifying as a master cabinetmaker. After spending 10 years practicing his craft, he studied wood science and technology at the University of Hamburg and obtained a doctorate in wood science. Since 1999, Dr. Lukowsky has worked at the Fraunhofer Institute for Wood Research (WKI) in Brunswick, Germany, where he specializes in failure analysis and wood protection.

Contents

Acknowledgments

I should express my deep gratitude to my esteemed colleague Anja Lütte for her critical review of the manuscript and her many suggestions. Many of the methods described in the following pages are the product of our joint efforts. This book would never have been possible without her support.

I am likewise indebted to Martin Ohlmeyer and Gerald Koch from the Thünen Institute of Wood Research in Hamburg and my colleagues Peter Meinlschmidt, Norbert Rüther, and Andreas Zillessen at the Fraunhofer Institute for Wood Research (WKI) in Braunschweig. Their invaluable assistance with numerous technical issues helped me to avoid several pitfalls.

Many other colleagues, staff members at other institutes, students, and friends provided advice and help with the book. I would particularly like to thank Peter Böttcher, Lucile Wendt, Katharina Turhal, Sandra Hofmeister, Richard Lagemann, Kirsten Wittenberg, Ute Schliephake, Pablo Guindos, Barry Goodell, Grant Kirker, Rusty Swindoll, Tim Rypstra, Tunga Salthammer, Jeff Lowinski, Tekla Krebs, Melanie Torenz, and Sonja Steinmann.

I also wish to thank my dear father, Hans Lukowsky, not only for his wholehearted support, but also for his great efforts in teaching me critical thinking.

Needless to say, any remaining errors in this book are my own.

All critical comments and suggestions regarding additional analysis methods or possible improvements to the book are welcome. Please feel free to send me an e-mail at dirk@lukowsky.net.

Failure Analysis of Wood and Wood-Based Products

CHAPTER 1

Introduction

Wood and wood-based products are time-honored building materials and those properties that are critical to construction practice are well understood. In any case, minor manufacturing irregularities and tough service conditions are regularly tolerated by assemblies of wood and wood-based products without any notable failures. The standard bonding and coating techniques for wood and wood-based products also have a proven track record and, in most countries, are specified in standards or codes. Significant failures are accordingly rare.

Moreover, in most cases where failures involving wood and wood-based products do occur, their causes can be conclusively established on site through the application of common sense, logic, and technical knowledge. Key factors in this regard include the distribution of failures, the time of failure occurrence, and the ambient and service conditions. Failure investigators typically need to check for compliance with the relevant regulatory documents (standards, codes, guidance papers) during materials manufacture, processing, or installation. This may sometimes require the assistance of specialist laboratories.

The number of failures where the cause is not readily identifiable, where deep-seated causes need to be investigated or where no clear-cut guidance exists is relatively small. These are the cases primarily addressed by the methods and strategies described in this book.

Failure analyses typically have to be conducted under cost and/or time pressure. The effort required for the analysis has to remain in reasonable proportion to the financial loss. Less elaborate analysis methods with an (assumed) high probability of success are therefore preferred where the amount of the loss is low. Conversely, in the case of very high losses or personal injury, for example, it may be appropriate to apply complex techniques even where, in the investigator's experience, the discovery of useful insights, though not impossible, remains unlikely. In the case of failures in industrial manufacturing processes or during construction work, the prompt delivery of results is often crucial. These circumstances also call for methods that combine maximum speed and reliability. However, despite the cost and time pressure, all investigations must be reproducible and observe the relevant scientific principles.

Straightforward quick-check methods frequently offer a suitable means of clarifying failures or, at least, developing a framework for the application of more specialized techniques. The use of methods simple enough to be performed during the site visit or immediately upon the receipt of specimens may eliminate the need for more extensive analyses. One aim of this book is to present these relatively uncomplicated methods and to show that, for instance, the combination of sandpaper up to grit P1000, a low-cost USB microscope, iodine solution, acute observation skills, and technical knowledge, is all that is needed to conduct in-depth, professional investigations, also directly on site. Attention is also drawn to those cases where tests by specialist laboratories using elaborate equipment offer the only means of pinpointing the failure causes.

The level of detail required of the results may vary widely from case to case. Sometimes, the sole purpose of the analysis is to confirm or rule out a particular cause. In other instances, the only aim may be to identify the party at fault. Frequently, however, a determination of the precise failure cause is required, possibly in conjunction with proposals for remedial action. The nature of the available background information is also subject to wide variations. On the one hand, the investigations may center on a well-documented industrial manufacturing process that is inspected on site in the presence of the production managers, who—due to their immediate interest in an objective explanation of the causes—are eager to assist. At the other extreme are cases where specimens are provided by the potential party at fault, with no record of the sample-taking process nor any other background details or with consciously or unconsciously biased information. In any event, a healthy degree of skepticism vis-à-vis the available data is absolutely essential for a professional analysis.

A further challenge, on top of the failure analysis itself, relates to the presentation of the findings and conclusions. As a general rule, the report must be geared to professionals (e.g., expert consultants) while remaining intelligible for parties with no in-depth technical knowledge (e.g., attorneys and business persons). The various means of achieving this include written documents, photos, sketches, textbook references, and specially prepared explanations of the technical relationships or terminology. The approach recommended in this book is illustrated by graphic, easily comprehensible examples which, as far as possible, dispense with technical language.

The field of wood analysis includes a number of heavily specialized areas that can only be touched on here. These include structural timber design, the identification of fungi and insects, measurement and appraisal of emissions, identification of wood species, wood grading, detection of wood preservatives, as well as all associated legal issues.

The question as to whether deviations from the specified condition of a product constitute a defect is beyond the scope of this book. Even within Germany, for instance, some defects are evaluated differently in the north and south. Similarly, opinions about what constitutes a defect-free product tend to differ widely from country to country. This may be attributed to the varying climatic conditions and, probably even more so, cultural differences. No consideration is therefore given here to the quality requirements specified by national and international standards, codes, or guidance papers.

CHAPTER 2

Investigation Strategy

The process of elimination is a common approach in failure analysis. The actual failure cause is identified by systematically narrowing down the possible explanations. Subsequent investigations can then be limited to answering the remaining questions and to confirming or disproving the failure hypothesis.

Every cause that can be ruled out with certainty is knowledge gained. It is a common human weakness that supposedly "negative" cognitions are less appreciated than "positive," confirming cognitions. Yet, for the purpose of gaining a better understanding and of helping the client, the elimination of possible failure causes is often just as valuable as circumstantial evidence (indications) of the actual cause. Moreover, when failure causes cannot be determined with certainty, discounted causes can at least reduce the scope of the possible explanations.

Parallel findings (see Sec. 2.3) are an invaluable aid in ruling out failure causes. They help to determine whether failures are independent of spatial, temporal, material-related, or production-related differences. If parquet installed in both rooms with and without underfloor heating shows signs of damage, then the underfloor heating system can be ruled out as the solitary cause. If failures occur both at the edges and within the surface of a furniture front, the surface grinder can be ruled out as the sole cause of the damage. And if discolorations occur not only on exposed surfaces but also under furniture handles, UV rays can be discounted as a relevant factor. Upon closer inspection, however, even the above interpretations of parallel findings are still founded on a number of assumptions. Take the last example: "The occurrence of discolorations under furniture handles rules out UV rays as the actual cause." This is based, among others, on the following premises:

P1: The fitting has been in place for a long time or since the item was delivered.
P2: A possible manipulation of the color of the surface under the fitting (e.g., with the intent to deceive the investigator) has been ruled out.

P3: There was no possibility that the UV rays hitting the surface could have led to the formation of substances that penetrated under the fitting and caused the same discoloration there.

P4: There was no possibility that the surface under the handle could have been specially treated prior to installation and that this treatment caused the same discoloration as the UV rays hitting the surface.

P5: There was no possibility that a chemical reaction between the fitting and underlying surface could have caused the same discoloration as the UV rays hitting the surface.

Premise P1 can be confirmed or disproved by relevant enquiries. It can also be safely assumed that a real-life door will not be used without fittings.

Any investigator can be tricked into reaching false conclusions by clever, intentional manipulation. As a rule, premise P2 is accepted as given and no further investigations are undertaken in this regard.

For logical reasons, P3, P4, and P5 are accepted as givens at the start of the investigation as experience shows the probability of such unusual circumstances to be very low. Only where all other possible causes have been ruled out in the course of the analysis might it be useful to scrutinize these premises (see example in Fig. 4.18).

In an expert opinion, it is unhelpful to present too many different sets of conclusions based on competing premises. It is the investigator's task, based on his or her special knowledge, to accept particular premises (i.e., make certain assumptions) and perform further targeted investigations on the basis of these.

2.1 Avoiding Pitfalls

The outcome of failure analyses frequently has far-reaching financial consequences for the parties involved. Failure analyses also serve as the basis for changing procedures, materials, or constructions to prevent future failures. Anyone conducting such analyses thus assumes considerable responsibility, which makes it vital to avoid errors. Yet, as the saying goes, "to err is human." In the field of medical diagnostics, surveys have been conducted to establish how frequently errors occur:

> *Despite enormous advances in the diagnostic options, particularly in respect of technology and instrumentation, the few existing studies undertaken in recent years and decades on the subject of misdiagnosis suggest—with a single exception—that the misdiagnosis rate in large hospitals and university clinics in North America and central Europe remains flat at around 10%.* [48]

The equivalent figure for wood analyses is unknown. There are contradictory arguments for both a higher and a lower misdiagnosis rate in failure investigations on wood and wood-based products than in the healthcare sector. My personal knowledge of numerous reports suggests that 10% is by no means wholly implausible. Figure 2.1 identifies the typical errors that cause doctors to misdiagnose a patient's condition. While some of the factors are undoubtedly limited to the medical profession, others are also relevant to the field of materials science. While doctors may be less interested in the causes of an illness than in the cure, for the investigator establishing the causes may be an indispensable step in producing a "diagnosis." Both failure investigators and doctors sometimes have the tendency not to take the complaints of their "patients" seriously and to reduce the problem, for example, to the hypercritical attitudes or behavior of the users. On the other hand, unlike normal medical examinations, expert opinions on wood failures are frequently directed at the work of colleagues or potential customers: one

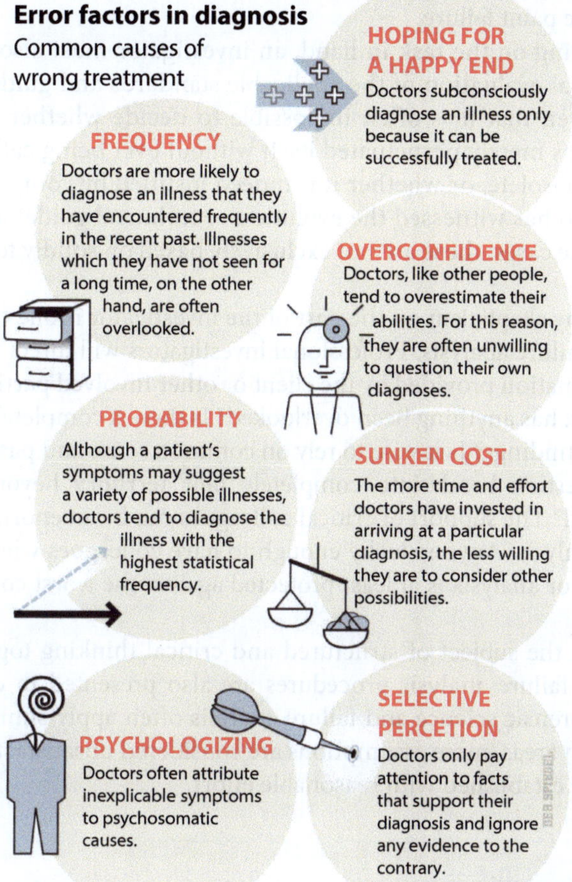

Figure 2.1 Typical errors potentially leading to misdiagnosis in the medical profession (*Der Spiegel* magazine, 7/2011).

master carpenter judges the efforts of another; a wood technologist evaluates the products of a wood-based panel plant, etc.

While wood or wood-based panel failures may have a single cause, the actual situation is frequently more complex. The eventual failure may result from a combination of adverse factors, parallel defects and more or less unknown influences. Should a defect be identified in the course of an investigation, it is thus essential to establish whether this really is the root cause.

For some investigators, the only criteria for judging a defect are the requirements set out in standards, guidelines, and contracts: if, in the case of flaking paint on a window, the film thickness is found to be only 80 µm instead of the specified 100 µm (0.1 mm), the investigation—so it may seem—need not proceed further given the obvious deviation from the requirement. Yet, although the inadequate film thickness may well have technical and legal implications, it can only be considered as one possible cause among many for the paint failure.

Depending on the task in hand, an investigator must also be prepared to undertake a critical evaluation of the applicable standards and guidelines. This tends to be difficult given that it is often impossible to decide whether a particular regulatory provision has merely perpetuated itself without ever being called into question and is effectively obsolete, or whether it is indeed justified by compelling technical reasons. Anyone who has witnessed the evolution of technical guidelines knows only too well that, in some cases, these are not exclusively based on solidly founded, state-of-the-art knowledge.

A healthy skepticism on the part of the investigator is one of the absolute essentials for expert failure analysis. Professional investigators will direct this skepticism not only at the information provided by the client or other involved parties, but also at their own assessments: has anything been overlooked? Is there a completely different way of interpreting the findings? Is it wise to rely on common sense and past experience or does the case in question break into completely new territory beyond anything previously encountered? The support of critical colleagues can be an enormous help in sharpening a failure analysis. Anyone lucky enough to have colleagues who are willing and able to dissect a poor analysis is at least protected against the worst consequences of misinterpretation.

Tips on the subject of structured and critical thinking together with information relevant to failure analysis procedures are also presented in criminalistics textbooks [18, 73]. Forensic science and failure analysis often apply similar methods and strategies. In both areas, mere assumptions are considered unsatisfactory in cases where hard facts can be established with reasonable effort.

2.2 Analysis Specimens

In some cases, a decision has to be reached even during the initial telephone enquiry on the nature and quantity of specimens needed for analysis or the necessity of a site visit. Likewise, in the course of a site visit, a decision may need to be made on whether to take samples for analysis and, if so, how many. Both in determining the type and quantity of specimens to be submitted and in extracting samples during a site visit, crucial decisions have to be taken on the basis of incomplete information.

It is virtually always useful, if not essential, to take samples from both damaged and failure-free areas. Furthermore, components exhibiting failures should also be cut up in such a way as to include failure-free areas (should such areas exist). This is because important clues may already have been obliterated at the failure location and that the inclusion of undamaged areas may cast light on the distribution of failures. Particularly useful are additional specimens that were either not installed or installed under significantly different circumstances. In many cases, it is precisely the samples without any (discernible) failures that play a decisive role in identifying causes. These samples sometimes offer the only means of assessing the basic suitability of a material for its designated use. To minimize the number of samples needed as much information as possible should be gathered in advance, by means of telephone interviews or preliminary investigations during the site visit. The example in [24] lists 76 questions on moisture-induced veneer failures that illustrate the full spectrum of potentially relevant information. A similar list of 65 questions on the possible causes of parquet flooring failures is presented in [62].

The following considerations may influence the decision as to the nature of the samples to be submitted:

- What questions need answering?
- Is the description of the facts coherent?
- Are there any photos showing the failure in its context?
- Is there a single, clear-cut failure pattern?
- Are the areas exhibiting failures clearly demarcated from failure-free areas or are there flowing transitions?
- Are there any areas where external actions (use/climate) can be largely ruled out, for example, below hardware?
- Is the component particularly valuable and will it be destroyed during the analysis?
- How big are the components?
- Does any potential failure cause immediately present itself?
- Do any other materials also exhibit failures?
- Is it possible to submit additional samples if needed or would a further site visit, for example, be needed?

2.3 Parallel Findings

The term "parallel findings" is used here to denote all phenomena encountered during the analysis that indicate differences in component or material behavior in response to ambient or manufacturing conditions, or differences in failure distribution or chronology. Key questions include:

- Where are failures present, where are they absent, and what is the difference between these locations?
- When did the failures occur and when were they not present?
- To what loads/exposure are the relevant components subjected? What about the failure-free components?
- What production parameters were used in manufacturing the damaged components? What production parameters were adopted for the failure-free components?

Parallel findings may be obtained by visual inspection or may only result from a more detailed, for example, microscopic, examination. In many analyses, parallel findings ultimately deliver the decisive clues in clarifying the failure causes.

2.3.1 Distribution of Features

The most important step in developing the framework for an analysis, especially at the start of the investigation, is to examine the distribution of features. This procedure is illustrated in Sec. 4.2. Distribution can be considered at various levels: within the building, on particular objects and on individual components.

Typical questions include:

- Are other materials in close vicinity damaged or undamaged? In case of extreme moisture exposure or drying out, for instance, would other materials be expected to show similar signs of corrosion or cracking?
- Do the service conditions of the damaged and undamaged components/component surfaces differ in any way?
- Does the failure distribution depend on compass direction?
- Do failures also occur at edges or on covered areas (e.g., below hardware or carpets)?
- Is the same failure distribution pattern observable on several components?
- Is there any correlation between the distribution of features and certain stages of production?
- Do features appear both along and across the grain?
- Are heartwood and sapwood equally affected?
- Do the features run parallel and/or perpendicular to the production or grain direction?
- Do the failures exhibit any characteristic pattern?

2.3.2 Chronology

The chronology of failures may sometimes be an important starting point for the investigation (see Chap. 4). On the one hand, the precise (absolute) time of failure may need to be determined (did the crack appear before or after April 1, 2014, the day of the transfer of risk to the buyer?). On the other hand, the absolute or relative chronology maybe the key to identifying the failure cause.

Knowledge of the time at which failures initially occur provides important clues about the failure cause. All factors, production steps and service conditions that postdate the first recorded observation of a failure can be safely ruled out as the root cause.

In some cases, the production period may offer a first indication of the potential failure cause. In colder climatic zones, the air in woodworking plants tends to be very dry in winter, with relative humidity of 20% being anything but uncommon. Products or components manufactured under such conditions may well show signs of swelling during service. Extremely dry air can also prevent the proper curing of reactive polyurethane adhesives. By the same token, excessively low humidity during manufacturing can normally be ruled out as the failure cause, where material batches from both summer and winter production are equally affected.

Typical questions include:

- When exactly was the failure noticed and can the possibility be ruled out that it was previously overlooked?

- When/over what period were the relevant products manufactured?
- When were the upstream products (adhesive, wood-based panels, coating) delivered? Were they supplied in several batches?
- During what period and under what environmental conditions were the components shipped and stored?

2.3.3 Organizational Factors

Batches are not always produced in a single operation. Nor is an entire batch necessarily installed at only one location. Organizational differences such as these may sometimes offer a source of parallel findings.

Typical questions include:

- Were other failure-free products manufactured at the same time as the components in question? Do any undelivered materials remain from the original production run?
- Were there any follow-up deliveries?
- Were the materials also incorporated at other locations?
- Do simultaneously delivered, though differently produced materials exhibit similar failures?
- Were any suppliers switched?
- Did similar problems occur with other manufacturers' products?

2.3.4 Comparison with Standard Commercial Materials

The broader an investigator's experience, the greater the likelihood of a "spontaneous" assessment as to whether a finding is normal or exceptional. In unusual cases, and particularly with microscopic material analyses, however, comparisons with standard commercial products are often needed to reach a conclusion. Figure 2.2 shows how the

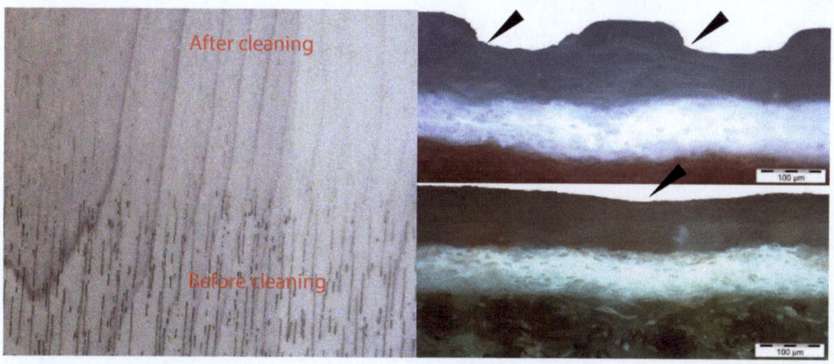

FIGURE 2.2 Cleaning of the dirty laminate floor surface (left) necessitated excessive time and physical effort. The deep, almost vertical contours of the embossed finish (micrograph of cross-section top right) are very difficult to clean. As a comparison with ten standard products showed, the embossing on laminate is normally much shallower and has rounded angles (bottom right), which makes it far easier to clean.

critical features of a dirt-prone laminate are identified through comparison with the embossed pore texture of several standard laminate products.

Comparative investigations may also be needed to decide whether a defect actually exists. In the absence of clear-cut regulations, it may be appropriate to class as defective any product whose performance is significantly inferior to that of most standard products on the market (from all price categories).

CHAPTER 3

Methods

The methods and strategies presented below have stood the test of time and practice. Some of them can be applied with minimal effort and without any expensive special equipment. Indeed, experience has shown that simple methods—quite apart from being fast and economical to implement—can often deliver results that are unobtainable using standard laboratory techniques. Naturally enough, there are still cases where elaborate analytical methods, such as FTIR spectroscopy, energy-dispersive X-ray (EDX) spectroscopy, or high-performance thermal-flow thermography, are indispensable. Such investigations can be performed by specialized laboratories, as needed, and are also discussed here for the sake of completeness.

It is not possible to know, in advance, which investigation methods will be most useful in any particular case. This will ultimately depend on factors such as time pressure and the amount of material available. Procedures often have to be adapted creatively in line with specific applications. Hence, the following compendium of methods is also designed to provide inspiration for the investigator's own experimentation.

3.1 Documentation

3.1.1 Photography

Photos are an indispensable part of failure analyses. They are used to document evidence and eliminate the need for elaborate written descriptions, serve as a memory aid, and can sometimes even be regarded as an investigation method in their own right. However, many surface failures either cannot be recorded in images or can only be captured by laborious procedures. Small surface elevations, cracks, or color nuances that are relatively easy to discern with the naked eye can sometimes be near-impossible to catch in a photo. One of the reasons for this is that the human eye generates three-dimensional pictures while photos are two-dimensional. Also, our brain pieces together the representation of an object from several different images, based on the different perspectives and/or lighting conditions that result from the movement of the onlooker or the object itself.

Suitable options for making failures visible include oblique illumination, the utilization of reflections, extreme magnification, and certain forms of image editing (Figs. 3.1 and 3.2). Investigators should nonetheless remember that viewing objects in overly bright lighting conditions is not a suitable way of judging whether certain features actually constitute

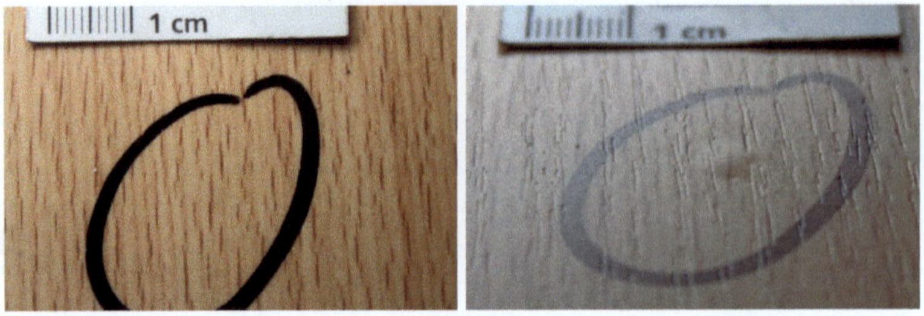

Figure 3.1 The surface elevation on a melamine-faced board can only be recorded under oblique lighting.

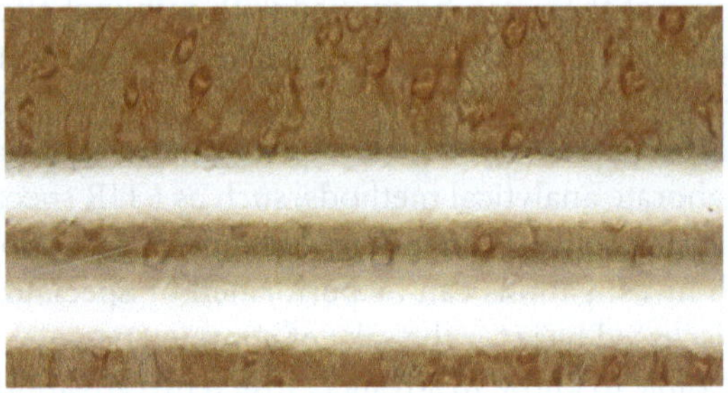

Figure 3.2 On shiny surfaces, even minute elevations measuring only a few thousandths of a millimeter can be recognized and documented along the boundaries of the reflection produced by fluorescent lamps.

a "defect." Such assessments should always be performed under normal illumination and from a normal viewing distance. However, where the aim is to establish and document failure causes—which is the focus of this book—the above qualifications are irrelevant and all means which serve this end are legitimate.

Any form of image editing is acceptable if it helps to establish the truth. Many fine structures cannot be distinguished at all without increased image sharpness or contrast on printouts (Fig. 3.3). Also, with shots taken during site visits, the lighting conditions are often so poor that the relevant details would be unrecognizable without image editing. Purists may rightly object that this opens the door to blatant manipulation, given that relatively minor details can be grotesquely exaggerated. Yet, the very same problem arises—in fact, to an even more serious extent due to the less obvious "manipulation"—in the case of close-up photos and, particularly, micrographs. The choice of picture detail and microscopic analysis method offers enormous scope for an investigator with manipulative intent to turn any finding into its opposite. Viewed under the microscope, virtually any surface will show irregularities, which can be made to seem problematic under a high enough magnification. If investigators cannot be expected to choose representative details, then no analysis can be trusted. It is equally important to be confident that the adopted image editing technique will not suppress any relevant details. To dispel any doubts, the report can present the original and the processed pictures side by

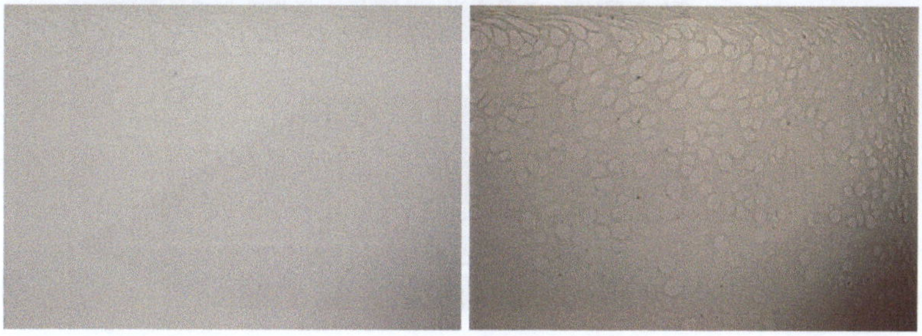

Figure 3.3 Coating defect before and after image editing.

side. However, even this practice can be deceptive: although it appears to offer absolute objectivity, it has no bearing on the key sources of manipulation, that is, the choice of picture details and lighting conditions.

Here are a few basic suggestions regarding photos:

- Wide-angle lenses distort the contours of objects. Although taking shots from a greater distance with a telephoto (zoom) lens reduces the line distortion, it simultaneously increases the risk of camera shake.
- The quality of photos is improved by a plain background and a straight view of the object.
- To obtain a good image, several shots should always be taken under various lighting conditions and from different perspectives. Three levels of detail—overview, detail and close-up—are often useful for photographic documentation.
- The inclusion of human-scale objects makes photos easier to comprehend (Fig. 3.4).

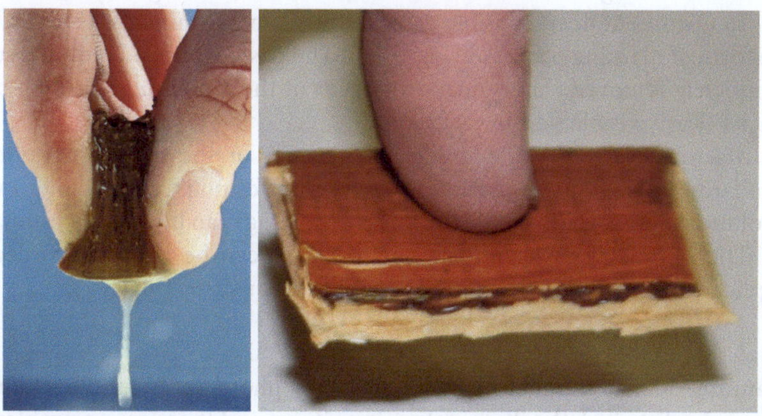

Figure 3.4 The left-hand photo indicates the degree of softening and saturation of the wood: water can be squeezed by hand from the specimen, which has been destroyed by bacteria (after centuries under water). The right-hand photo shows the tackiness of a coating: the specimen sticks to the finger and can be lifted by adhesive strength alone.

FIGURE 3.5 Surface elevations are observable as dark zones below the blade of a backlit knife-edge straight-edge.

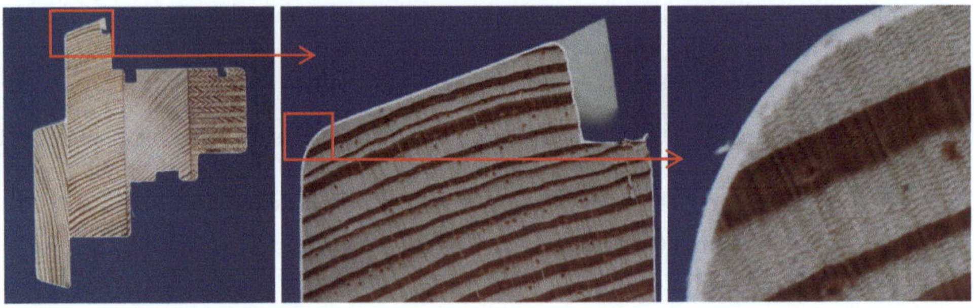

FIGURE 3.6 Polished section of a window scantling: overview and detail shot with a macro lens plus magnified close-up. The white coating is approximately 100 μm (0.1 mm) thick.

With a knife-edge straight-edge (a rule with a sharp measuring edge), it is possible to document height differences down to around 2 μm (Fig. 3.5).

To document near-invisible cracks, these can be stained with dyes, for example, in the form of marking paste (see Fig. 3.29). The paste is rubbed onto the surface and then completely removed. The contrasting color of the remaining marking paste makes the cracks clearly visible on the surface.

Macro lenses on single-lens reflex cameras or the macro mode feature of many standard cameras can be used to take close-ups of polished specimens with impressive resolutions (Fig. 3.6). Even the thickness of coatings can be roughly estimated in this way.

The symmetry or asymmetry of materials can be clearly recorded by creating a duplicate mirror image of cross-section photos or scans (Fig. 3.7).

3.1.2 Scanning

Scanning offers a number of advantages over photography: the contours of specimens remain undistorted, the illumination is uniform and there are no problems with camera shake. On the other hand, scans are restricted to flat surfaces and the resolutions achievable by ordinary scanners are lower than those of shots taken with macro lenses. In some cases, scanning is almost the only means of recording minimal height differences

FIGURE 3.7 An asymmetrical five-layer plywood construction led to severe deformation. To demonstrate the asymmetry, a macrophoto of the cross-section was duplicated and mirrored horizontally and vertically using a standard word-processor with integrated image editing tools. Had the layup been symmetrical, the layers of both the original and mirror image would lie in the same plane.

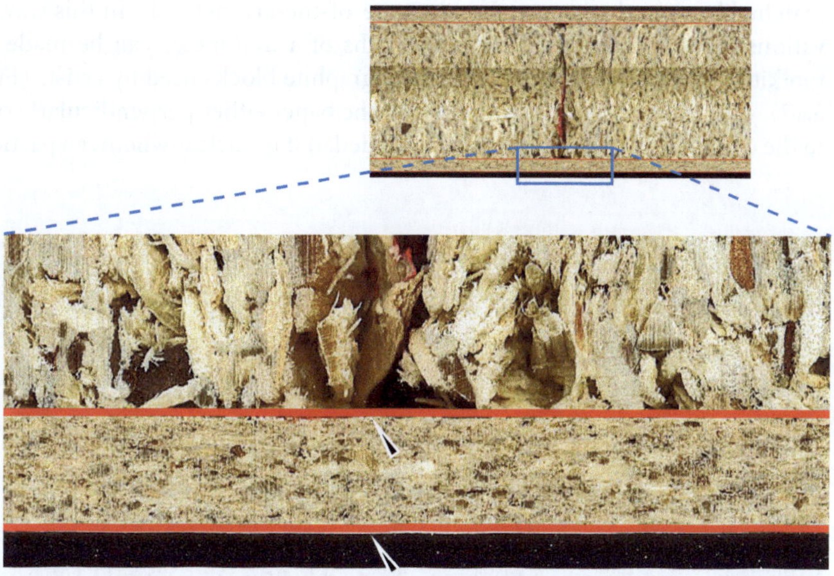

FIGURE 3.8 Door cross-section with magnified close-up documented with a standard office scanner. A slight swelling is observable in the core layer while the thin surface layer shows no change in thickness. Standard graphics programs or even normal word-processors can be used for the exact positioning of lines along surfaces or layer boundaries.

on larger cross-sections. Although photos allow greater lengths to be captured in high resolution, the slight degree of distortion inherent in photos make them unsuitable for this type of application. Figure 3.8 shows the scan of a door cross-section. By means of magnification, it was possible to trace minute signs of swelling at the joint in the core layer.

3.1.3 Replicas

In rare instances, a direct investigation of specimens may not be possible, for example, because destruction of the relevant objects is unacceptable or too great an effort would be involved. One option here is to make molded replicas of the surface texture. Replicas also offer a useful solution in cases where a precise record of the original situation is required prior to the dismantling of hardware or other components. Dental casting materials lend themselves to this purpose, as do certain compounds used for DIY. The two casting components are kneaded together in the ratio 1:1 and pressed onto the surface. The casts fully cure within a matter of minutes. Impressions can also be taken on damp surfaces. The fact that the cured materials are easy to slice with a cutter or scalpel makes it relatively simple to document the relevant cross-sections by micrographs (Fig. 3.9).

3.1.4 Frottage

As a child, probably everyone has copied the face of coins by laying a piece of paper over the surface and rubbing with a pencil (Fig. 3.10).

For all its simplicity, the frottage technique can be used to produce blur-free images of miniscule elevations and depressions on areas of any size in a way that can hardly be matched by even the most elaborate, state-of-the-art methods. In this way, surface elevations measuring only a few thousandths of a millimeter can be made visible. The toolkit comprises normal paper and the graphite blocks used by artists (Figs. 3.11 and 3.12). The graphite block is rubbed over the paper either perpendicularly or diagonally to the direction of the textures to be recorded. If it is unclear whether a particular feature

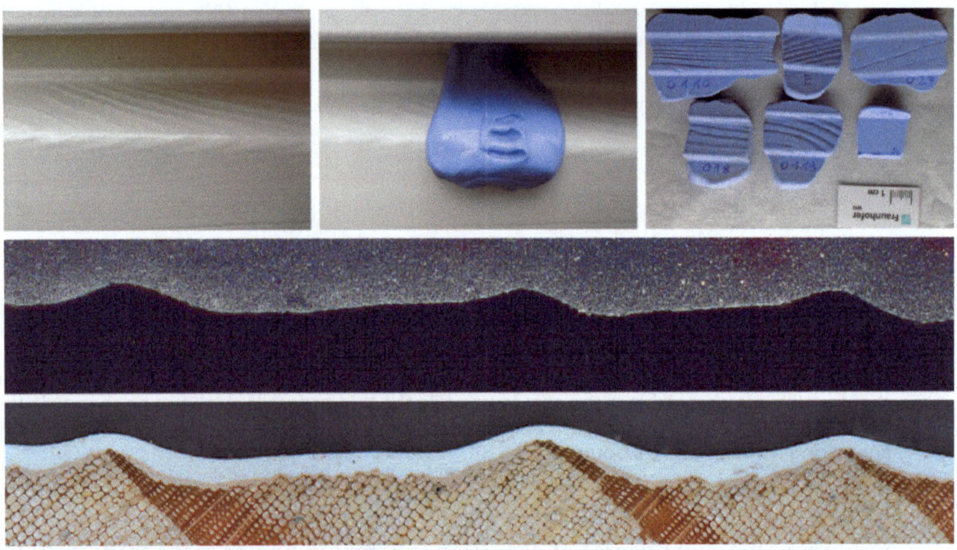

FIGURE 3.9 A number of replicas were produced as a nondestructive record of the surface texture of painted wood members during a site visit. The bottom-right picture shows a cross-section of one of the affected components, together with the associated replica above. As the replica is at the top of the micrograph, the black background shows the exact surface contours.

Figure 3.10 Frottage during boring school lessons.

Figure 3.11 A graphite block is rubbed across a piece of paper placed over the specimen to document its surface texture.

Figure 3.12 The approximately 20-µm-deep features on a high-pressure laminate (HPL) surface, which are practically impossible to capture in a photo, were recorded by placing paper over a large area and rubbing with a graphite block. Some of the features that emerged had not been previously detected, even by close inspection under oblique illumination. The frottage indicates that, after the application of the adhesive, the underside of the panel had been pushed sideways while travelling over the disk rollers of the conveyor. This had caused some of the adhesive to be rubbed off.

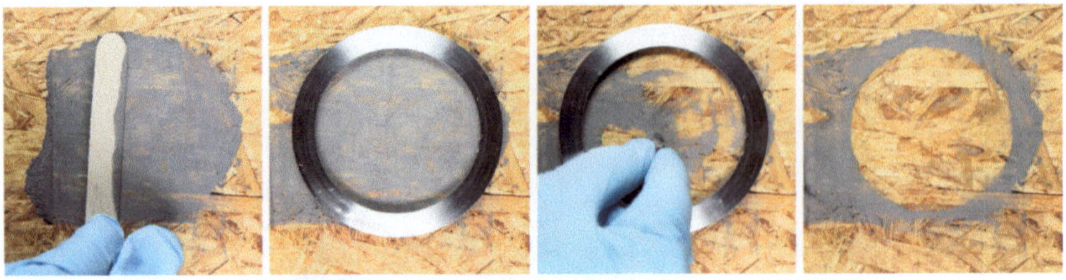

FIGURE 3.13 Determination of surface roughness of OSB (oriented strand board) panel. The iron dust is applied to the surface and removed by magnet from a predefined area inside an aluminum ring.

really exists or has been produced by accidentally tilting the graphite block, the frottage should be repeated in a different direction. To capture small-scale depressions, the object can also be rubbed with an obliquely held pencil.

3.1.5 Estimation of Surface Roughness

Surface roughness can be approximately quantified by establishing the volume needed to even out the irregularities. The finer the anticipated irregularities, the finer the material to be applied. In the simplest case, the panel surface is smoothed out with a filler and the quantity of material needed per unit area (application rate) determined. However, the results obtained here may also be influenced by panel suction.

To eliminate the impact of suction in determining surface roughness, panels can also be coated with powder. Iron dust is ideal for this purpose in that it is heavy and allows full, controlled removal from the surface by magnet. After ascertaining the mass of iron dust removed from a predefined surface area, the volume of the surface pores can be calculated using the previously established powder density (Fig. 3.13).

3.1.6 Fluorescence

Fluorescent substances are able to absorb high-energy UV radiation that is invisible to the human eye and reemit this at a lower-energy and thus visible wavelength. Radiation sources exist for various wavelength ranges. For the purposes of wood analyses, the simple torches used in some countries to check banknotes are frequently adequate. They emit light in the wavelength range roughly between 350 and 390 nm (so-called "black light").

Even better results can be achieved using special UV LED torches (Fig. 3.14). These torches emit light with only a single wavelength, for example, 365 nm, and thus provide a better contrast. The better-quality torches also offer an extremely high UV intensity and are even capable of revealing features that remain practically invisible under normal black light. The staining described in Sec. 5.1.4, for example, was only discernible using a powerful LED torch with a wavelength of 365 nm. It was not detectable under black light.

Certain materials are known to exhibit fluorescence at these wavelengths. For example, papers treated with optical brighteners glow a bright white under black light. The heartwood of black locust or honey locust, on the other hand, shows up in an intense

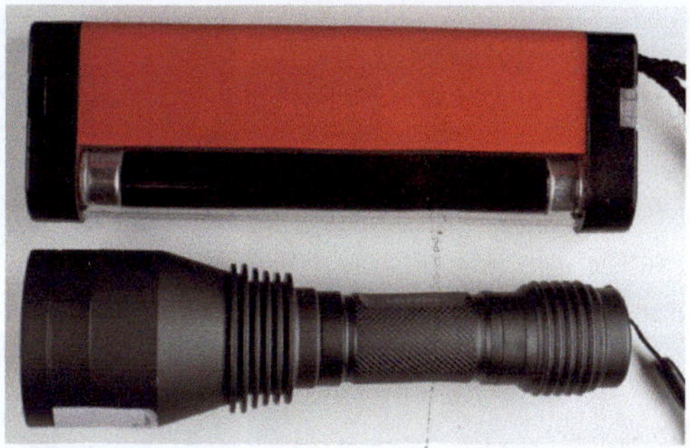

FIGURE 3.14 Black-light banknote checker (top) and UV LED torch (bottom).

FIGURE 3.15 Left: Under a black-light lamp, freshly cut honey locust heartwood glows an intense yellowish-green. Right: Inspection under black light reveals particles of black locust in a particleboard panel. Flakes of this kind are, of course, absolutely acceptable in this type of product. Such findings may nonetheless be useful in specific cases in checking, for example, whether all the panels under investigation originated from the same batch.

yellowish-green (Fig. 3.15). Other substances that only sometimes show fluorescence, for example, many vegetable oils, cannot be reliably detected. In numerous cases, it is not possible to predict whether or not a material will fluoresce. Fluorescence is generally caused by minute quantities of contaminants or additives, which may or may not be present in any particular specimen. Consequently, the method can only indicate differences between specific locations. For an analysis of failure causes, however, this may prove decisive (Figs. 3.15–3.17).

A special trick can be used to obtain a uniform image of the fluorescence on a larger surface using the weak light of a simple UV torch (Fig. 3.18). Here, the camera must be mounted on a tripod and the exposure time set to 30 or so seconds. During the exposure, the UV torch is then moved rapidly back and forth over the specimen.

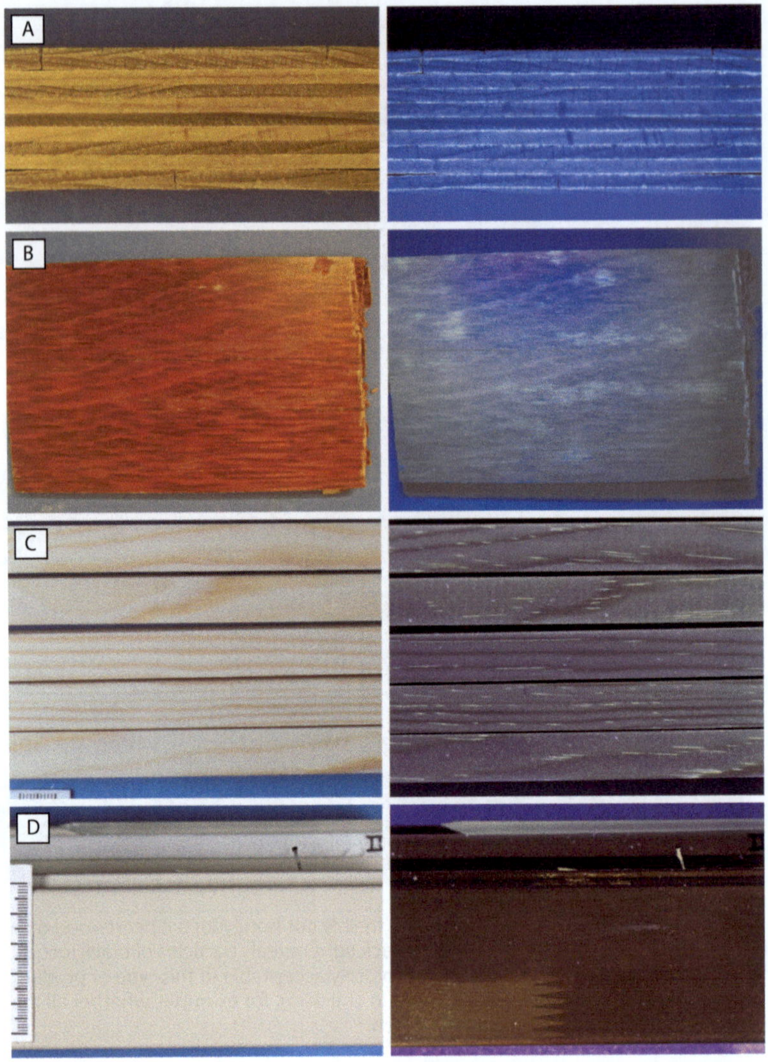

FIGURE 3.16 The adhesive in a plywood panel (A) and the distribution of troublesome heartwood extractives in a jatoba parquet product (B) are revealed under the black light of a banknote checker. Examples C and D show how the light of a powerful UV LED torch reveals the resin ducts in pine strips as well as heartwood extractives that have bled through the coating.

3.1.7 Accuracy of Measurement Results

Mathematical ignorance is by nothing else better reflected than in over-accurate presentation of results. —Carl Friedrich Gauß (1777–1855)

Statistical methods seldom play an explicit role in wood analyses. It is nonetheless vital for certain fundamental statistical principles to be observed in the evaluation of findings. Of crucial importance is an assessment of the representativeness of the investigated specimens and the expected error margin of the applied measurement method,

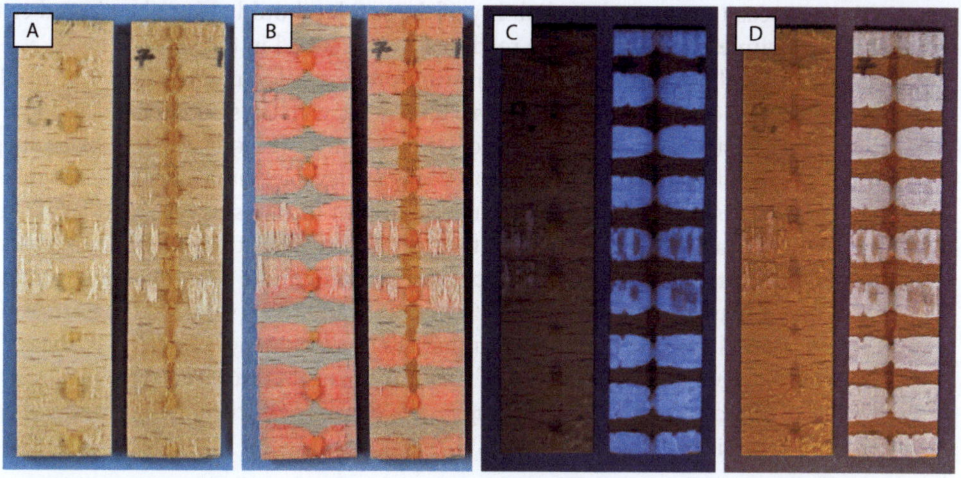

FIGURE 3.17 Two specimens of the debonded top layer of an engineered parquet product. Image A shows a normal view. The stain shown in image B was produced by the application of dimethylamino-cinnamaldehyde (DACA) and indicates the presence of polyurethane adhesive on both specimens. As the fluorescence images C (original) and D (edited) reveal, only the adhesive used for one of the top layers is observed to fluoresce. The adhesive material used for the parquet clearly originates from two different batches, only one of which contains a fluorescent substance.

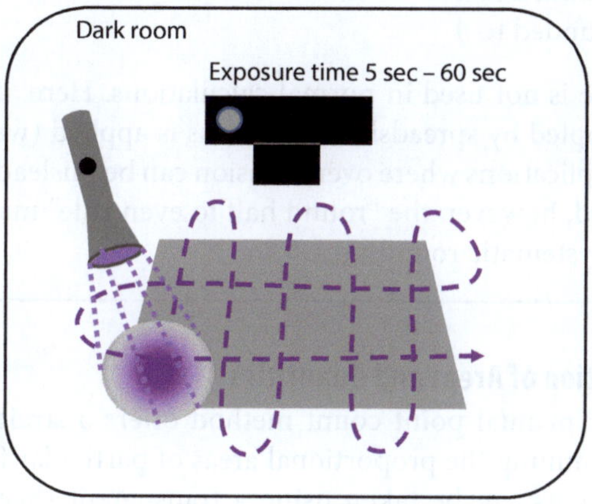

FIGURE 3.18 Fluorescence photography using the weak light of a simple UV torch.

both of which should be indicated in the report. Given that modern-day measuring instruments often deliver impressively accurate results, it is tempting to quote every last decimal place and trust the figures blindly. It is precisely because of the enormous confidence inspired by exact numbers that they should be viewed so critically. For even where the measurement method really does offer extreme accuracy, the relative degree of uncertainty introduced merely by the sample-taking process (e.g., time and place of specimen extraction) will normally be so great as to justify the rounding of measurement readings as a better reflection of the inherent confidence level. Particularly with electronic moisture measurements, the reliability of the results should always be viewed critically, with the assumed accuracy of the readings clearly stated in the report.

By way of illustration, length-measuring programs for micrographs can display the thickness of wood coatings to two decimal places. Yet, with the standard display of the results in micrometers (symbol "µm," equal to 1/1,000 mm), even one decimal place suggests a greater degree of accuracy than is theoretically achievable with optical microscopes, even under perfect conditions. The determination of film thicknesses on wood substrates is in any case never particularly exact due to the material's cellular structure. For this reason, even µm readings without any decimal places can never be more than qualified estimates. To reflect this, it may often be appropriate to round the indicated values to the nearest 10 µm.

As a general rule for numeric data that are not particularly exact, it is always better, for instance, to write "2 m" instead of "2.0 m." This is because the zero in the first decimal place, especially for laypeople, suggests a degree of accuracy that may not actually exist and is often unnecessary.

For anyone in doubt as to whether to round 1.5 to 1 or 2, or 4.5 to 4 or 5: to minimize rounding errors, the "round half to even" rule is recommended:

1.5 is rounded to 2
2.5 is rounded to 2
3.5 is rounded to 4
4.5 is rounded to 4

This rule is not used in normal calculations. Here, the standard "round half up" rule adopted by spreadsheet programs is applied (with 2.5 rounded to 3 and not 2). In applications where overprecision can be misleading and rigorous rounding is applied, however, the "round half to even rule" may be a more appropriate method of systematic rounding.

3.1.8 Estimation of Areas and Quantities

The systematic manual point count method offers a straightforward, though precise means of determining the proportional areas of particular features. Even on site, accurate measurements can be taken using a transparent sheet with a printed grid. The method described in ASTM E562-08, for example, involves the use of a grid to quantify the occurrence of particular features at certain points. If sufficient points are counted, the proportional area covered by the various features can be statistically estimated.

The systematic manual point count method makes use of a gridded transparent sheet, which is placed over the object or image under investigation. With digital photos, a similar grid can be electronically superimposed on the screen image. For each grid cell, the feature located, for example, at the bottom-left intersection point is identified. The ratio of the number of points counted for the relevant feature to the total number of points gives the overall proportion of the surface area exhibiting the feature. The number of points counted for the relevant feature multiplied by the area of one grid equals the absolute area occupied by the feature.

The systematic manual point count method can be used with any type of surface pattern. Particularly where the transitions between features are slightly blurred, for example, as between earlywood and latewood (Fig. 3.19) or with discolorations, this technique can—in conjunction with the necessary technical knowledge—deliver a reasonable estimate of proportional areas. While image analysis programs are able, in such cases, to obtain results that appear very precise, this precision is ultimately founded on subjective decisions as to where exactly the boundaries lie (Fig. 3.20).

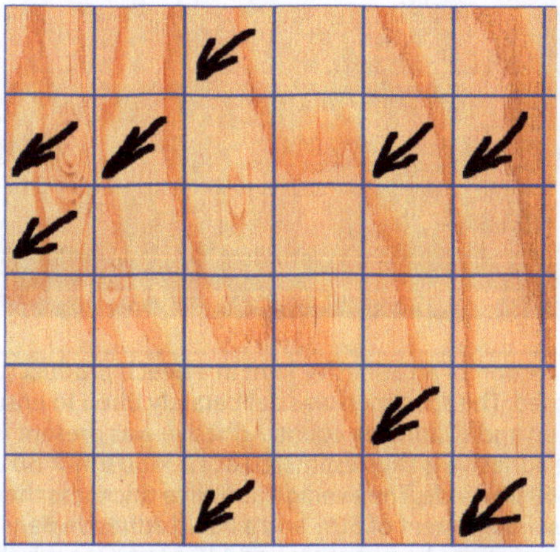

FIGURE 3.19 Latewood is identified in the bottom-left corner of nine of the 36 grid cells. This estimate thus puts the proportion of latewood on the surface at 25%. One wrongly marked cell would result in an error of around 3%.

FIGURE 3.20 To facilitate analysis, the picture is converted into a black-and-white image on the basis of precise specifications for classifying the individual color channels. This analysis puts the proportion of latewood on the surface at 31.52%. Yet, for purely wood anatomical reasons, it is evident that some of the features were wrongly identified as latewood by the analysis. The exactness of the numeric value should not trick the reader into thinking that the result is anything more than an estimate.

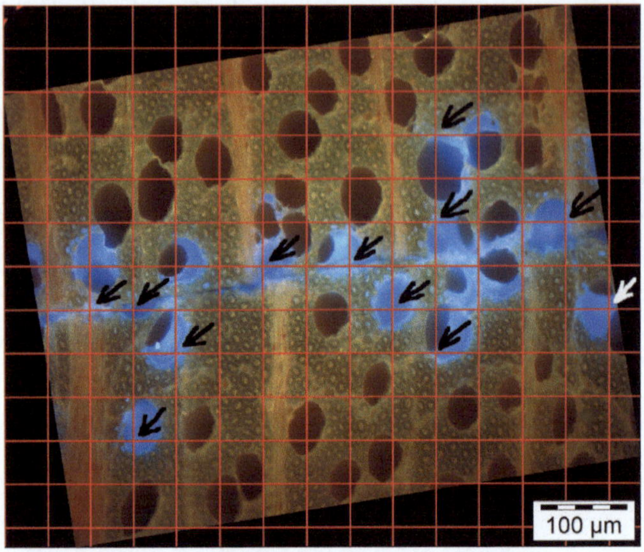

Figure 3.21 Cross-section of an adhesive joint in a beech product (UV fluorescence). A grid with 50 µm cells was superimposed. The specimen was deliberately tilted to ensure that the adhesive joint does not run parallel to the grid and distort the count. Adhesive was present in the bottom-left corner of 12 cells. The total area of the adhesive is thus estimated at $12 \times 50\ \mu m \times 50\ \mu m = 30{,}000\ \mu m^2$. The examined wood specimen is 660 µm wide. The average adhesive thickness therefore totals $30{,}000\ \mu m^2/660\ \mu m = 45\ \mu m$. Assuming a solids content of 50% and a liquid adhesive density of 1.2 g/cm^3, the application rate of the liquid adhesive can be put at around 110 g/m^2. One misallocated grid cell would make a difference of 8 g/m^2 in the estimated liquid adhesive quantity.

Figure 3.21 illustrates the application of the systematic manual point count method to microscopic examinations, in this case to estimate the quantity of adhesive in a joint. The approximative nature of the result is due less to the possibility of measurement errors than to the inadmissibility of treating a 660-µm (0.66-mm) specimen section as representative for an adhesive joint. Should highly reliable results be needed, the same procedure would have to be repeated at ten or so further locations.

3.2 Sample-Taking

Normally, clear-cut statements on failure causes can only be made on the basis of destructive tests. Any case that can be unambiguously resolved without invasive methods must be viewed as a stroke of luck. As a general principle, however, destructive tests are essential for the achievement of reliable results. Samples are taken by sawing, drilling, cutting, or chiseling with sharp implements (Fig. 3.22).

At the same time, nondestructive investigation methods—in particular, a close observation of the failure characteristics and distribution—may substantially reduce the number of destructive tests required for the relevant components. Apart from thermography and ultrasonic coat thickness measurement, nondestructive methods also include relatively simple operations such as the careful rubbing and tapping of surfaces. Here, the resulting sounds can sometimes reveal a lot about the substrate.

FIGURE 3.22 Sample-taking tools.

3.2.1 Systematic versus Random Sampling

Where only the failure cause needs to be determined, the sample-taking process is straightforward and can, for example, be based on the criteria set out in Sec. 2.1. In such cases, investigators will systematically extract specimens on the basis of their knowledge, experience, and judgment. On the other hand, should the incidence of failures also need to be established in cases where a full investigation of all components/surfaces is not feasible (e.g., for efficiency reasons or to minimize damage to the existing material), the method and intensity of sampling becomes a central issue. Randomness of selection is then the foremost principle, that is, each location, component, etc., must have the same probability of being investigated.

This necessitates the development and strict application of a sampling, investigation, or documentation regime (Fig. 3.23). Unsystematic sample-taking without such a

FIGURE 3.23 Sample-taking from a package of beech strips to establish milling quality. The black numbers on the strip ends indicate the specimens selected for investigation under the adopted sampling regime.

plan will not generate a truly random set of specimens. A custom-designed plan is required for each particular case. The numerous possible selection methods range from the investigation of every 10th component in an assembly or the random choice of specimens from lists and construction drawings to the rolling of dice.

Many national standards specify detailed criteria and sampling regimes for assessing the conformity of sawn timber. For a lot comprising nine packages and a total of 1,000 items, for instance, European prestandard prEN 12169 (2006) requires testing, in all, of 80 specimens taken from three of the packages. In case of doubt, the strategies described there can also be adapted to other specimen types.

3.2.2 Drilling Cores

Drilling cores can be extracted using hand-operated or mechanical drills. Machine coring, in particular, is a fast and efficient method of producing large, continuous specimens without any risk of mechanical damage during sample-taking (Figs. 3.24 and 3.25). While the density of the drilling cores can be reliably determined, the moisture content may be significantly influenced by the friction heat generated during the extraction process, especially if the specimens are very damp: the smaller the drilling core, the blunter the

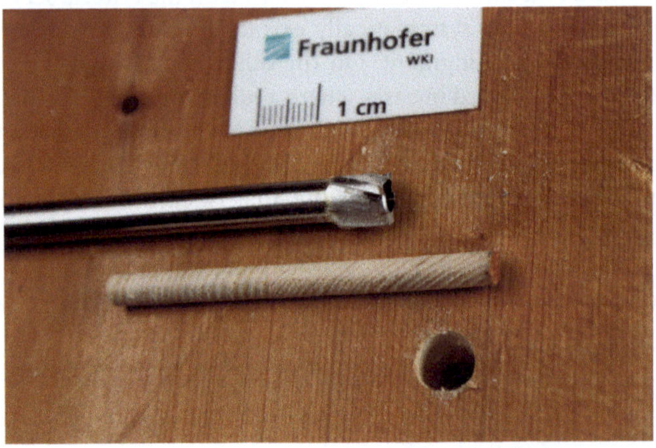

FIGURE 3.24 Special drills allow thin cores to be extracted from specimens.

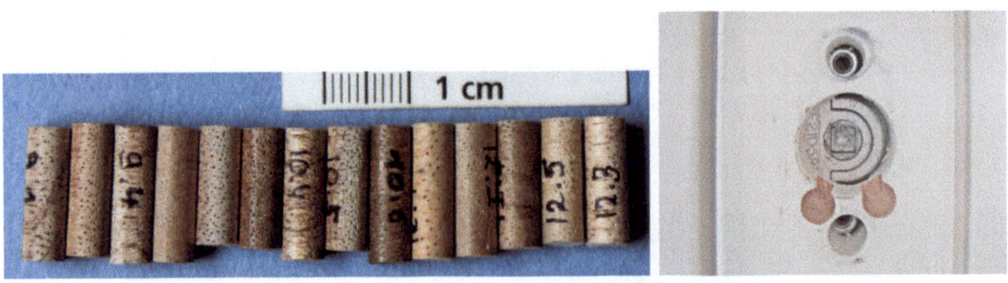

FIGURE 3.25 Drilling cores taken from surface below window handle plates to determine wood species and density as well as the coating thickness.

drill and the faster the speed, the greater the deviation from the actual value. For a more reliable analysis of the results, then, parallel tests should always be conducted on woods of known moisture content.

3.3 Sanding Down

Knowing whether or not the failure on the component surface corresponds to particular features occurring in lower layers can sometimes provide decisive clues as to the failure cause. The reason for some surface failures may become immediately apparent after carefully sanding away the top layer. The particular advantage of this method—in contrast to the microscopic examination of cross-sections, which only ever yields information about minute portions of a specimen—is that it allows the investigation and integral documentation of large surfaces with relatively little effort.

Surfaces can be sanded down manually or with a wide-belt sanding machine. As mechanical sanding allows the appearance of a specimen to be recorded at predefined levels, it can deliver the exact vertical profile of a feature. In practical applications, however, manual sanding usually suffices.

It is often crucial to map the exact position of the original surface defects onto the sanded substrate. This can be achieved by one of the following methods:

- Photos taken from the same position before and after sanding.
- Recording of the points for investigation by tracing them onto a transparent sheet or glass. The advantage here is that the position of markings can be superimposed on the specimen at any required time during the investigation (Fig. 3.26).
- Small drill holes adjacent to the failure location that extends down to the substrate (Fig. 3.27).

FIGURE 3.26 The cracks on the melamine coated particleboard were traced onto a transparent sheet prior to sanding. This allowed the original crack path to be superimposed onto the sanded panel.

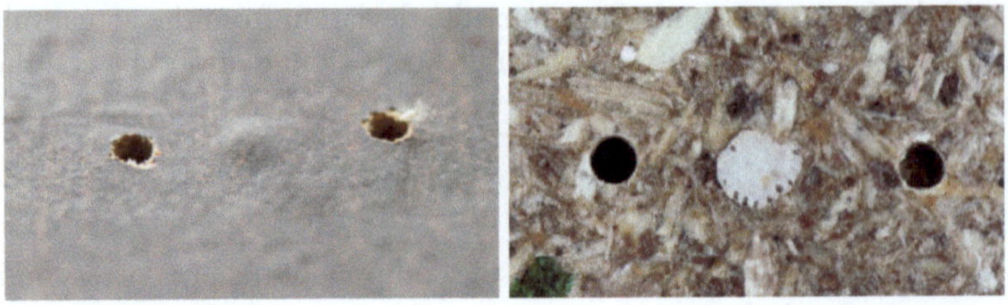

FIGURE 3.27 The position of a small hump on coated particleboard was marked by two 1-mm-diameter drill holes. Sanding down of the coating revealed a white plastic particle between the holes.

FIGURE 3.28 Unsightly elevations were discovered on the surface of a veneered particleboard panel. The beech veneer was carefully sanded down to reveal uncompressed adhesive beads. Between these beads there was no adhesive layer to protect the panel against infiltration of the wood stain. The stain was thus able to reach the surface layer of the particleboard, where it caused local swelling.

Figure 3.28 shows how the sanding down technique was used to trace the cause of blistering on beech-veneered particleboard to uncompressed adhesive beads. Wood stain penetrated through the adhesive-free zones down to the particleboard and caused local swelling.

The images in Fig. 3.29 show the surface of a specimen with coating cracks, which was sanded down to different levels. Prior to sanding, the cracks had been stained with marking paste and the surface with a blue dye. The cross-sectional sketches in Fig. 3.30 show how the images resulted. Further examples of the sanding down technique are presented in Figs. 3.127 and 4.16.

It may sometimes be useful to sand down surfaces in such a way as to produce a tapered profile with simultaneously visible individual layers or coats spread out over a wide area. These can then be inspected and further analyzed even without a microscope. To obtain a relatively well-defined taper angle, a strip of adhesive tape is stuck to the sandpaper. One end of the specimens remains in contact with this strip during

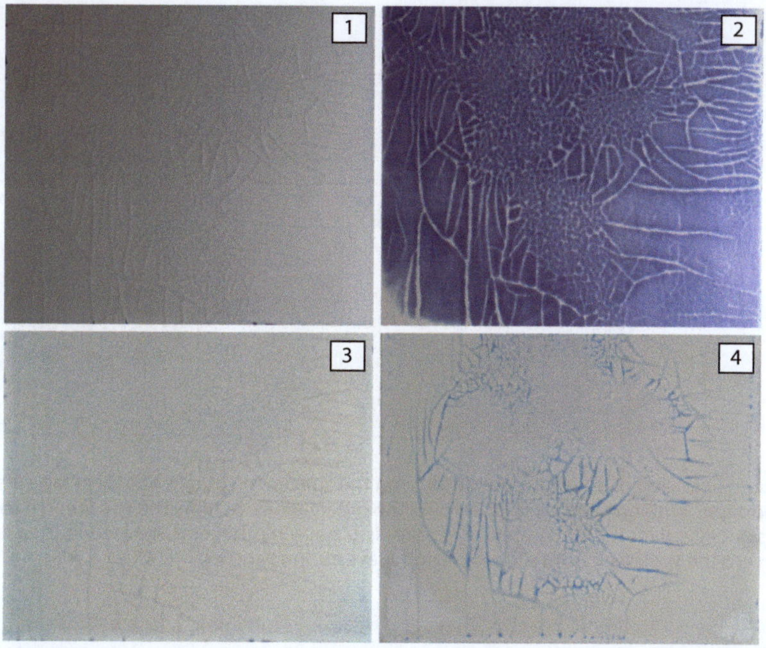

FIGURE 3.29 Cracks in a furniture panel finish. The cracks were first stained with marking paste and the panel surface with the dye cotton blue (normal fountain pen ink would also have been suitable). During sanding, white zones first appeared along the edges of the cracks, as seen in Picture 2. The fact that the blue stain was removed here first indicates that the crack faces were arched slightly upwards. Although this was detectable by touching the panel, the sanding down technique provided a clear record. The cracks became less visible as sanding progressed (Picture 3), with the stained crack width providing the only color contrast. Shortly before the primer paper was reached, the large cracks then became wider (Picture 4). This was due to the marking paste, which had penetrated into the debonded zones at the crack base and was visible through the wafer-thin coating. The cross-sections in Fig. 3.30 show the evolution of the images.

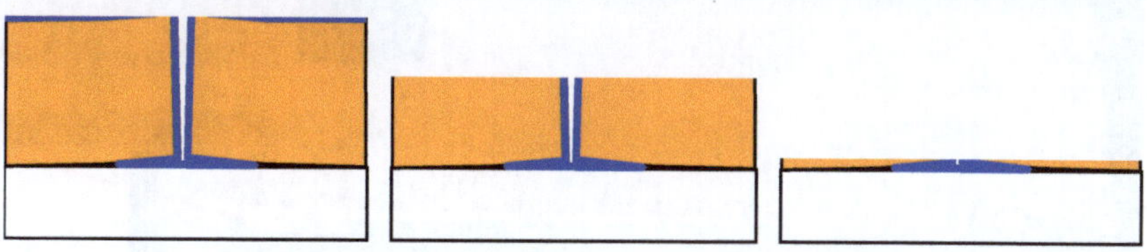

FIGURE 3.30 Cross-sectional sketches showing how the images 2, 3, and 4 in Fig. 3.29 resulted.

sanding and is not abraded (Fig. 3.31). Depending on the thickness of the layer under investigation and specimen length, it is possible to produce tapers down to an angle of around 0.5°. This would, for example, allow a layer originally 100 μm thick to be spread over a width of around 5 mm—equivalent to a 50-fold magnification. Where both the substrate and taper are perfectly even, it is possible to determine the thickness ratio between the various layers (Fig. 3.32). The levelness of the sanded surface can be checked using a knife-edge straight-edge.

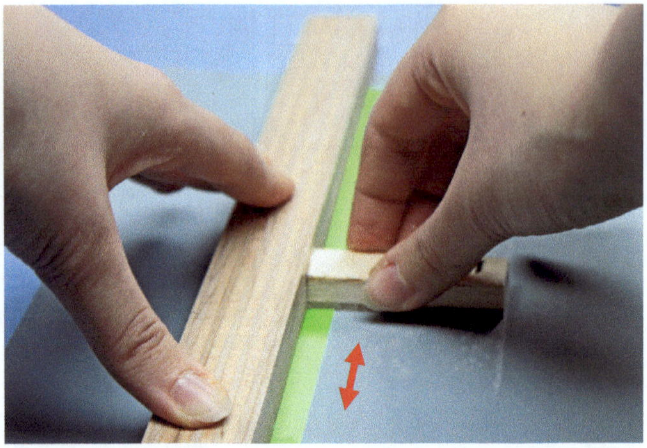

FIGURE 3.31 To produce a tapered profile on the specimen surface, adhesive tape is stuck to the sandpaper. With one end positioned over the green adhesive tape, the specimen is sanded for as long as is necessary for the substrate to become visible at the other end. A timber batten is used as a stop to prevent the specimen from slipping lengthways during sanding.

FIGURE 3.32 800-µm-thick (0.8-mm), four-coat finish, taper-sanded at an angle of 1°. Due to the taper, the sequence of coats is spread out over a width of 23 mm and available for further analysis. While manual sanding tends to produce heavier abrasion toward the edges, the thickness ratio between coats remains constant throughout. The coat thicknesses shown in the micrograph almost exactly correspond to the thicknesses determined by taper sanding.

FIGURE 3.33 At the interface between melamine paper and particleboard, gradual transitions between layers indicate a strong bond while abrupt transitions signify poor adhesion, for example, due to depressions in the panel surface.

Taper sanding involves the application of mechanical loads to layers. If these are very thin and do not properly adhere to the substrate, they will peel off during sanding. At the transition between melamine paper and particleboard, for example, this phenomenon will indicate which flakes formed a strong bond and which formed a weak bond with the paper (Fig. 3.33).

3.4 Simulation of Production or Service Conditions

In addition to the direct analysis of specimens, specially configured test set-ups (experiments) may also help to cast light on failure causes. Their usefulness, however, tends to be limited to relatively simple issues. Especially for complex industrial manufacturing processes, laboratory experiments are very elaborate and provide no guarantee of pinpointing the failure causes.

Among the experiments that may be conducted in the course of failure analyses are comparisons between different bonds and coatings. Typical questions that can be resolved in this way include the following:

- Is an appropriate product quality at all achievable with the particular adhesive/coating, even under optimum conditions?
- Is the same product quality achievable with retained samples as with other batches of the adhesive/coating?
- Does the adhesive/coating perform better or worse than comparable products on the market?

- Do the parameters presumed to be critical, for example, open time, substrate moisture content, application temperature, etc., have a major impact on the achievable product quality?
- Does the deliberate use of faulty production parameters (e.g., too low a pressure or overly high moisture content) lead to failures whose visual or microscopic characteristics resemble those of the investigated failure?

Only rarely do retained samples of both adhesive and adherends exist. Although the achievement of a sound bond between these samples under laboratory conditions may suggest an application error, it may nonetheless be wrong to assume that the test conditions are equivalent to those in the real-life manufacturing plant. Should it prove impossible to achieve a sound bond, further material combinations can be systematically investigated to establish the cause of the inadequate bond strength.

The sensitivity of components to specific ambient conditions can be determined by exposing them to conditions presumed to be still tolerable and, where appropriate, to conditions presumed to be critical. Any bond that fails under normal environmental conditions can be classed as defective. While this procedure appears fairly straightforward, such experiments may be exceedingly difficult to conduct in practice. Even the speed at which ambient conditions change may have a decisive impact on component behavior. Furthermore, a precise definition of the tolerable environmental conditions rarely exists (see Appendix A.2). Except in clear-cut cases, experiments of this kind tend to be more suitable for uncovering likely causes than for establishing hard facts.

3.5 Wood Moisture Content

Many failures are caused or influenced by moisture action and the exact determination of present moisture content is sometimes decisive. Still, of far greater importance in many cases is a reconstruction of the moisture values prevailing at a particular time in the past. Typical questions include:

- What was the moisture content of the materials during production?
- What was the moisture content of the materials during installation?
- Is there any indication of the moisture conditions during service?
- Does the material exhibit any moisture gradient that indicates the direction, and thus the cause, of moisture penetration?

3.5.1 Electronic Measurement Methods

The available electronic methods for the estimation of wood moisture content by measuring electrical resistance or capacitance are fast, straightforward and largely or wholly nondestructive. The measurement accuracy, however, is relatively low and depends on a wide range of factors.

When investigators take moisture measurements with an electric moisture meter at a previously unknown site, the accuracy described in the informative appendix of

AS/NZS 1080.1 can serve as a first indicator for the reliability of the results (reproduced with permission from SAI Global Ltd. under Licence 1310-c118):

Resistance moisture meters *provide reliable results in the moisture content range 7% to 22%, provided species and temperature corrections are made. Such meters can be used above 22%, but their accuracy diminishes as moisture content increases to such an extent that they are generally considered unsuitable for use above 40%. They are also unsuitable for measurements below 6% moisture content. At moisture contents between 10% and 12% accuracy is greatest at approximately ±0.5% to ±1% moisture content. Above 12% moisture content a variance of ±1.5% to ±3% moisture content can be expected with higher density species generally of lower accuracy.*

Hand-held capacitance moisture meters *in the moisture content range 7% to 22% have accuracies between ±3% and ±6% moisture content. The accuracy is very dependent on density and board thickness.*

Electronic moisture content readings are sometimes heavily influenced by salt deposits, wood preservative treatments or adhesives, with the displayed figures normally being too high. As the presence of such materials can never be ruled out in failure analysis, the accuracy of electronic moisture measurements is substantially improved by means of calibration against the oven-dry method for the tested material. The derived accuracy depends on the homogeneity of the material and the number of cross-checked samples.

Devices with insulated electrodes allow determination of the moisture content at various depths (Fig. 3.34). Without insulated electrodes, the maximum moisture level encountered between the electrodes is always displayed.

The results of electronic moisture measurements should be presented without any decimal places as these would suggest an unattainable degree of accuracy. And for all electronic moisture measurements stated in a report, it should be clearly pointed out that these figures were displayed by the device and do not necessarily indicate the exact moisture content.

FIGURE 3.34 Estimation of moisture content with insulated drive-in electrodes.

3.5.2 Oven-Dry Method

Exact moisture contents can be determined by means of the oven-dry method (EN 13183-1, AS/NZS 1080.1, and ASTM D 4442-07). The sensitivity of the balance used to weigh the specimen must be a minimum of 0.1% of its nominal oven-dry mass. The specimen is then stored in an oven at approximately 103°C for several hours (usually between 6 and 24 hours, depending on the specimen size and wood species) and then reweighed in the oven-dry state.

$$\text{Moisture content \%} = \frac{\text{initial mass } g - \text{oven dry mass } g}{\text{oven dry mass } g} \times 100\%$$

The moisture content determined by the oven-dry method will be too low if the oven drying time is too short to evaporate all the water from the specimen. This can be prevented by weighing the dried specimen, returning it to the oven for, say, another two hours and then reweighing it. If the second weight is within 0.1% of the first, then the sample can be regarded as oven-dry. Under ASTM D 4442-07, the specimens should be cooled in a desiccator prior to weighing. The equivalent European standard requires weighing of the hot specimen.

The result delivered by the oven-dry method is the average moisture content of a specimen. It is, however, often useful to know whether moisture is uniformly distributed in the wood or whether a moisture gradient exists. To establish this, the specimen is first chopped into suitable pieces (Fig. 3.35). By determining the moisture content of the individual pieces, it is then possible to gain a picture of the moisture distribution within the original specimen.

The presence of residual coatings on the specimen, adhesives or large quantities of low-boiling-point resins (found in pines, for example) may influence the results of the oven-dry method. Factors relating to sample-taking and transportation should also be taken into account. Hence, the accuracy of the results obtained by this method should not be over-exaggerated. Particularly for smaller specimens, an assumed maximum accuracy of ±0.5% is recommended. ASTM D 4442-07 assumes the accuracy to be no greater than ±1% moisture content where the adopted procedure deviates from the standard's "Primary Method A."

FIGURE 3.35 Splitting of a window frame section to determine moisture gradient.

3.5.3 Geometric Estimation of Moisture Content

The degrees of swelling and shrinkage of a wood component remain constant over time. This means that the component's moisture content at the time of the last working operation can often be estimated from its size—subject to the following conditions:

- No restraint, that is, swelling and shrinkage are unobstructed
- No decay due to the action of fungi
- No extractive bleeding
- No inward migration of substances
- No long-term heating over approximately 90°C

In situations where unrestrained swelling and shrinkage are not possible—for example, with bonded flooring or well-glued mortise-and-tenon corner joints and between the individual layers of plywoods, multilayer boards, etc.—wood may be subject to plastic deformation. In the case of directly abutting wood components, the swelling pressure resulting from a moisture content increase of 3% or greater is likely to result in permanent deformation (crushing) [72]. When the component dries back to its original moisture content, it will be narrower than it was before swelling. This effect is illustrated in Fig. 3.36. Even though the parquet exhibited a uniform moisture content when the photo was taken, gaping joints mark the spot where the floor was exposed to moisture from a plant pot.

Given that end grain absorbs water much more readily than the other wood surfaces, the swelling here tends to be most pronounced. With parquet, this is manifested by typically wide joints at the ends of the parquet strips after moisture penetration (Fig. 3.37).

FIGURE 3.36 Wide joints mark the position where the—now dry—bamboo parquet was exposed to prolonged moisture seepage from a plant pot. The swelling caused the parquet strips to be squeezed together. After drying, the strips were narrower than before.

Conversely, severe water penetration can be ruled out in cases where the joints are relatively tight and where, given the totality of the circumstances (parquet type, wood species, bond with substrate, etc.), this feature would have been expected after moisture exposure. Here, it would be important to check whether the swelling pressure had been relieved by deformation of the adhesive. This would be signaled by a narrower joint between parquet and wall. Table 3.1 presents diagrammatically the features described by

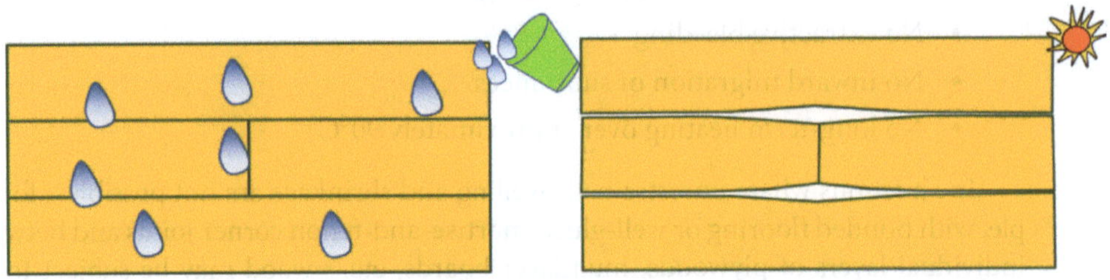

FIGURE 3.37 After water penetration, the plastic deformation of parquet strips is particularly pronounced at the end joints, where the greatest water absorption, and thus swelling, occurs. Joints such as these are a typical sign of plastic deformation. Uniform joints without any widening at the component ends mean that significant plastic deformation can be more or less ruled out.

Initial Moisture Content Too High	Moisture Content Too High During Service, Leading to Permanent Deformation	
	Of the Wood	**Of the Adhesive**
No movement of the parquet to the walls—unchanged gap between parquet and walls	No movement of the parquet to the walls—unchanged gap between parquet and walls	Movement of the parquet to the walls—smaller gap between parquet and walls
No typical signs of plastic deformation	Typical signs of plastic deformation	Few or no typical signs of plastic deformation
With checkerboard parquet, diagonally neighboring panels are precisely aligned in the longitudinal direction	With checkerboard parquet, diagonally neighboring panels are nearly aligned in the longitudinal direction	With checkerboard parquet, diagonally neighboring panels are not aligned in the longitudinal direction
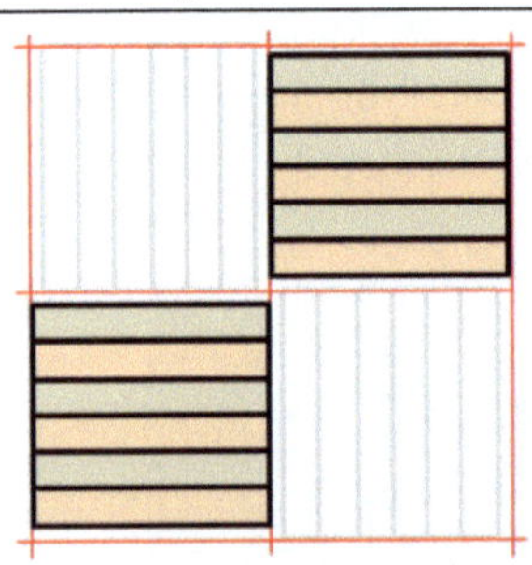	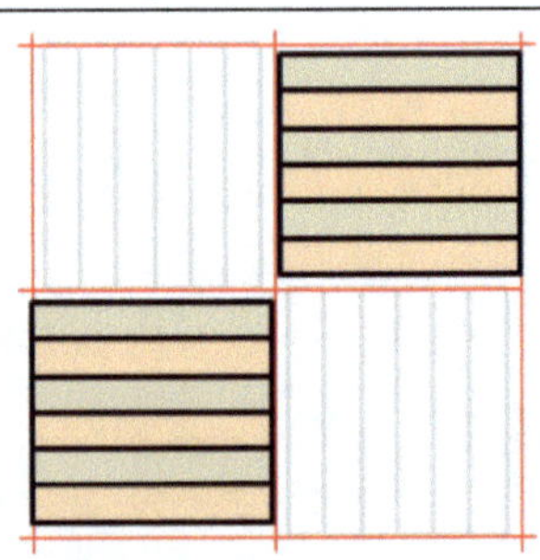	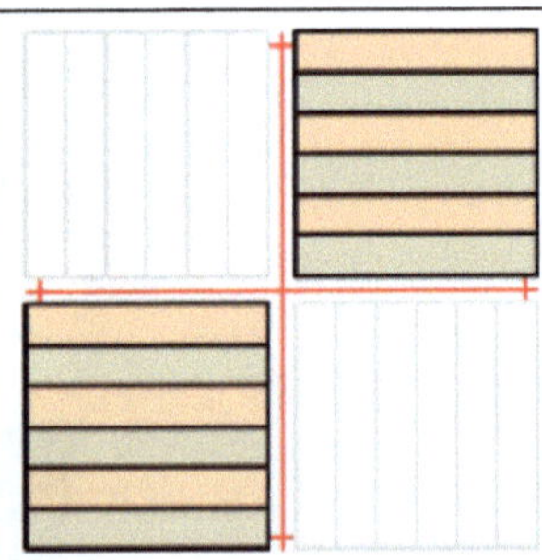
Panels and strips with smaller width than at time of assembly	Panels and strips with smaller width than at time of assembly	Panels and strips with same width as at time of assembly

TABLE 3.1 Pointers to the Different Causes of Large Gaps in Parquet (from Rapp [72])

Rapp [72] for simple drying, plastic deformation and adhesive displacement, based on the example of a checkerboard parquet floor.

If the exact size of a wood component after the last production step is known, then—provided it has been stored under restraint-free conditions—it is possible to estimate retrospectively the moisture content at the time of this last working operation using the characteristic swelling and shrinkage values known from literature (e.g., *Wood Handbook* [36]). In case of doubt, these values can also be determined for the installed wood in the laboratory.

As a rule, however, the manufacturer is the sole source of information on production sizes. Where disputes arise, then, such details are of limited use and must be treated with caution. Nonetheless, the moisture content during production can also be reconstructed if only the relative size after the last working operation is known. Estimates based on relative size make the (reliable) assumption that, under the known manufacturing conditions, a particular wood component will have the same length, width or thickness after the final production stage as the adjacent or mating component.

The corner joints of window sashes or solid-wood furniture are normally rabbeted or sanded down after bonding. Immediately after this operation, then, the long-pieces and cross-pieces are flush with each other. Should the moisture content of the components subsequently change, the extreme difference in swelling and shrinkage between long-pieces and cross-pieces will create an offset between these members. Although no offset would immediately result should the adhesive be able to withstand these loads, some signs of deformation would still be apparent at the joint. The presence or absence of this kind of offset can be extremely useful in establishing the moisture content of a component immediately after manufacture (Fig. 3.38). It is, however, essential to verify

Figure 3.38 Upon opening an overseas container, a number of small furniture items showed signs of heavy mold growth. The task was to check whether this was the result of inappropriate transport conditions. The long-pieces and cross-pieces were found to be flush at the corners of the table. The moisture content exceeded 30%. Traces of polyvinyl acetate (PVAc) adhesive on both long-pieces and cross-pieces, revealed by staining with Lugol's iodine, proved that the corners were sanded flush after bonding. The findings thus indicate that the wood also exhibited a moisture content exceeding the fiber saturation point of around 30% at the time the corner joints were glued. The most probable cause of the mold fungus growth was thus identified as an excessively high moisture content during production, rather than unsuitable transport conditions.

that the components really were flush after bonding. Where the window sash members are milled to the required profile prior to bonding and thus require no subsequent rabbeting, any offsets between long-pieces and cross-pieces cannot be used to deduce the original moisture content. At the same time, certain types of deformation in the components or cracks in the coating can still provide clues as to the moisture content at the time of production (Fig. 3.39).

To estimate the moisture content of veneers, laminates, etc., specimens of the relevant layers are first carefully extracted with a suitable tool, for example, a sharp knife, depending on adhesive type and substrate. In other cases, a thin layer can be sawn off and the residue adhering to the back side can be sanded, scraped or planed away. The length of the specimen layer and that of the original bond are then compared (Fig. 3.40). As an alternative or in addition to this, the deformation of cut-free layers can also be assessed (see Sec. 3.6.5).

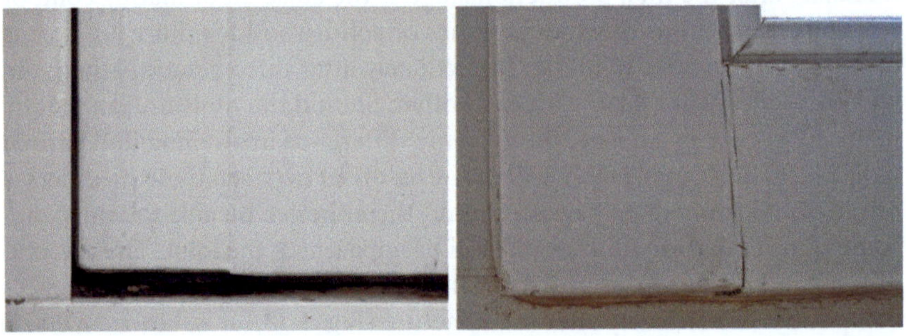

FIGURE 3.39 Offsets between the long-piece and cross-piece of window sashes. With the left-hand example, it must first be established whether the corner joint was at all flush after production. In the right-hand image, the deformation of wood and coating clearly indicates that the frame has swollen.

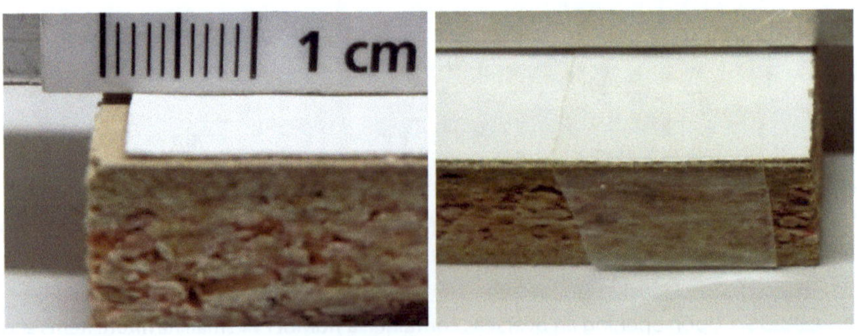

FIGURE 3.40 The cracked continuous-pressure laminate (CPL) on a particleboard panel was carefully cut away along the bondline and then laid back on top of the original panel. One end and the cracks were secured with adhesive tape. Over a 1-m length, the CPL was 2 mm shorter than the panel. This shows that the CPL had a substantially higher moisture content than the particleboard at the time it was bonded to the panel. Here, a combination of factors brought about the failure: a generously applied waterborne PVAc adhesive, a fire-resistant board product with low water absorption, the special pressing conditions plus an absorbent laminate with a relatively high degree of swelling.

While, at corner joints, the differential swelling and shrinkage of adjacent components ranges between 1:10 (axial : radial) and 1:20 (axial : tangential) and is therefore fairly easy to detect, the corresponding ratio for parallel strips or lamellas, for example, as in edge-glued panels, is at most 1:2 (radial : tangential). Yet, even here, in combination with the oven-dry method (see Sec. 3.5.2), important information can be obtained about the moisture content immediately after the final surface leveling operation. For this purpose, a graph is drawn with the thickness of the individual components plotted against the precise moisture content at the time they exhibited this thickness. Straight lines with a gradient reflecting the swelling coefficient (percentage dimensional change per percentage change in moisture content) applicable for the particular wood species and annual ring orientation are drawn through these points. If these lines intersect at a single point, then it can be assumed that this moisture content prevailed in all components at the time of the last surface leveling operation. Conversely, if the lines do not intersect at a single point, varying moisture contents can be assumed for the different components at the time of leveling (Fig. 3.41).

The example in Fig. 3.42 shows a window installed at a winter job site. Practically all window sashes were observed to have offsets at the corner joints on the internal face. The biggest offsets and highest moisture contents were always measured at the bottom edge of the sashes. This suggests that condensation had run down the panes and penetrated the corner joints. Despite this conclusive evidence, it was important to check whether the sash members may have been too dry at the time of manufacture. A minimum moisture content of 11% had been specified.

A glazing bead specimen was used to calibrate the electronic moisture measurements taken on site. By means of the electrical resistance method, the bead's moisture content was

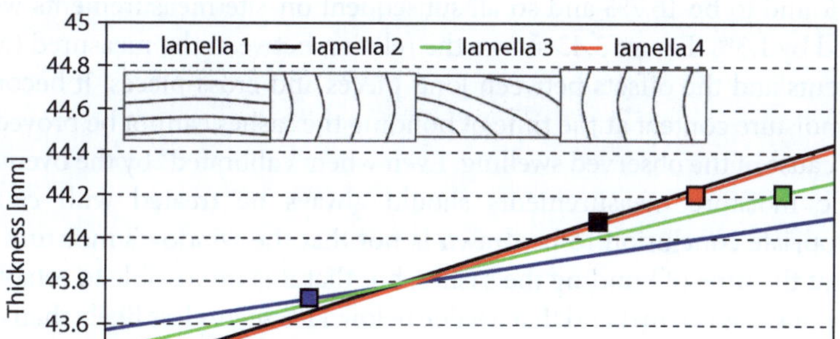

Figure 3.41 The squares indicate the measured thickness and moisture content for four lamellas in an edge-glued panel. The displayed lines running through the squares express the swelling coefficient for each lamella in function of the annual ring orientation (radial, tangential, 45°). The lines intersect at a moisture content of approximately 9.5%. This indicates that all lamellas exhibited a moisture content of between 9 and 10% at the time the panel was last sanded or planed.

FIGURE 3.42 Water running down the window panes during the winter construction period penetrated the corner joints of the sashes. The close-up (right) shows the offsets at the corner joints caused by swelling as well as the points at which the moisture content was measured. The overview (left) shows the mean values of the offsets measured for numerous windows in the property. The individual values are analyzed in Fig. 3.43.

estimated at 15.4% (mean value from four measurements). The oven-dry moisture content was found to be 16.7% and so all subsequent on-site measurements were upwardly corrected by 1.3%. Figure 3.43 shows the relation between the measured (adjusted) moisture contents and the offsets between long-pieces and cross-pieces. It becomes apparent that the moisture content at the time of bonding the sashes cannot be proved to be a contributory cause of the observed swelling. Even when "calibrated" by the oven-dry method, electronic moisture measurements should always be treated with caution. Hence, the appropriate conclusion to be drawn is not that the window's moisture content exceeded 11% at the time of bonding the sashes, but that it is impossible to establish that the specified value was not met and that a value below 11% seems less likely than a value above 11%.

A special case of geometric moisture content estimation relates to component wear. Evidence of wear/abrasion can be used to establish differences in levels in relation to adjacent components. This is illustrated in Fig. 3.44.

3.5.4 Arching

The swelling of wood floors is particularly likely to result in arching where the gap left between floor edge and wall is too small. Extreme arching failures are possible where end-grain wood block flooring is exposed to water leakages (Fig. 3.45). The considerable scale of arching those results from relatively small (constrained) changes in length, for example, due to swelling, can be vividly demonstrated using a folding meter: a 200-cm folding

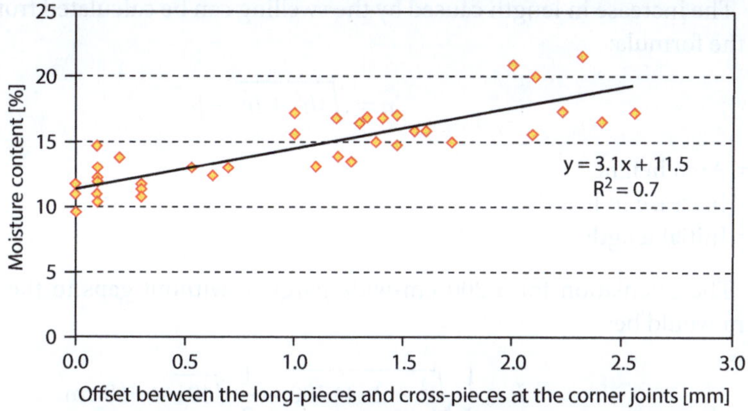

FIGURE 3.43 Moisture measurements plotted against offsets between long-pieces and cross-pieces at corner joints. Linear regression analysis shows a zero offset between the long-pieces and cross-pieces at a moisture content of 11.5%.

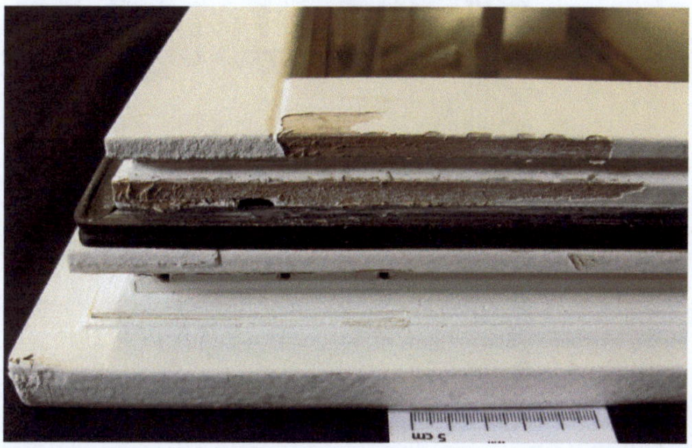

FIGURE 3.44 The bottom edge of a window sash shows signs of abrasion. The failure distribution indicates that the wood was swollen at the time the paint was worn away. Given that the remaining junctions between long-pieces and cross-pieces are flush in dry condition, it can be safely assumed that the junctions were also flush after manufacture. Yet, the fact that only the paint on the long-piece was worn away during use—which can only have occurred while the long-piece was swollen—implies that the window was heavily exposed to moisture for at least a short period during service.

meter arches to a height of 10 cm when placed between two abutments at a distance of 199 cm (Fig. 3.46).

The formula for calculating the arch height caused by constrained swelling is as follows:

$$h = \frac{1}{2}\sqrt{a^2 + 2ab}$$

The increase in length caused by the swelling can be calculated from the arch height by the formula:

$$a = \sqrt{4h^2 + b^2} - b$$

h = Arch height
a = Increase in length
b = Initial length

The calculation for a 200-cm-wide parquet, without gaps to the walls, that swells 1 cm would be:

$$h = \frac{1}{2}\sqrt{1^2 + 2 \cdot 1 \cdot 200} = \frac{1}{2}\sqrt{401} = \text{~}10 \text{ cm}$$

FIGURE 3.45 Water leaking from a defective valve flowed over this end-grain wood block floor for a whole day. As a result, the surface of the swollen wood arched to a height of approximately 50 cm.

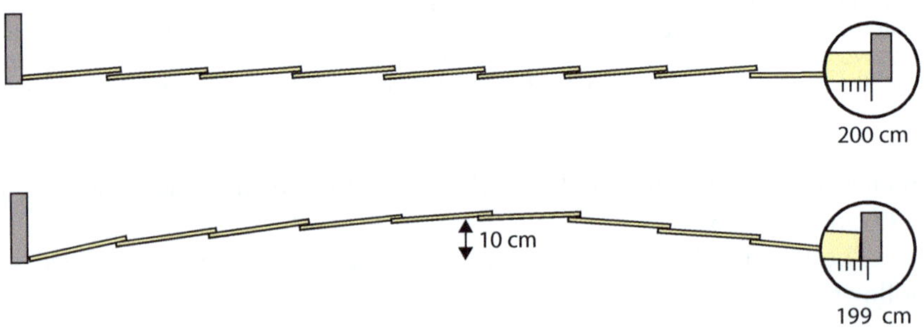

FIGURE 3.46 A 200-cm folding meter arches to a height of 10 cm when placed between two abutments at a distance of 199 cm.

Efflorescence on coated plywood

Wood extractives on a window coating

The peeling of the clear coat in the direction of the grain on the stair stringer immediately above the step was most likely caused by water on the upper surface of the step

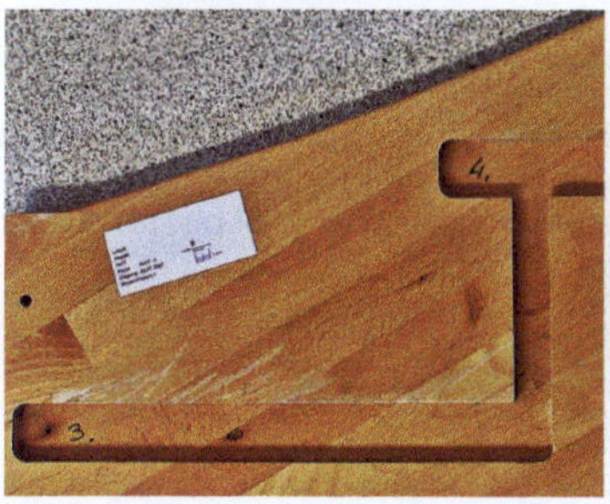

Mold fungi on a laminate floor

FIGURE 3.47 Tell-tale signs of moisture exposure during service.

3.5.5 Other Useful Facts about Moisture Exposure

Extractive bleeding, particular types of surface discoloration, efflorescence (see Sec. 5.10), characteristic changes to the surface of wood-based products, metal corrosion and mold fungi often provide clear evidence of water exposure at some point in the past (Fig. 3.47). At the same time, the lack of such clues does not automatically imply that no significant moisture penetration occurred. Tell-tale signs of moisture may be absent with woods that have no water-soluble extractives, wood-based products containing no efflorescence-producing salts, components treated with fungicidal preservative or hardware made of stainless steel. Yet, the situation is much clearer, for example, where both corroded and corrosion-free nails from the same material are present, as pictured in Fig. 3.48.

The degree of corrosion suffered by hardware, screws or nails depends on the wood species, temperature and moisture levels. This makes it difficult to draw general conclusions about the duration of moisture exposure on the basis of metal corrosion. Where necessary, tests can be conducted to determine the length of time needed to produce comparable discolorations for a certain level of moisture exposure.

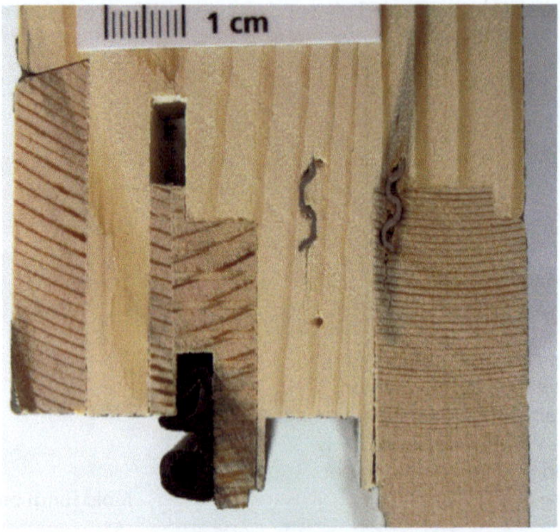

FIGURE 3.48 A corrugated fastener at a window corner joint (Scots pine) as the source of discoloration of the surrounding wood. Iron salt migration on this scale can be roughly estimated to take between several days and a few weeks, subject to a moisture content exceeding 30%. It is also clear in this example that the moisture originated from the interior (right-hand side of photo) given that the fastener nearer the outside shows no signs of corrosion.

3.6 Physical Investigation Methods

A discussion of the widely used test methods described in official standards is beyond the scope of this book. Failure analyses, however, often necessitate the performance of preliminary tests based on these methods. At the start of an analysis, at least, it is unimportant to establish whether the values specified by a particular standard have been met. A quick indication of basic suitability is normally enough to decide what further investigations are needed to clarify the failure. Moreover, setting up the environment needed for full-scale, standards-compliant tests is time-consuming and such tests often require more material than is available.

It is often possible to perform an initial, indicative physical test on bonds and coatings simply by scraping a surface with the fingernail. This may already suffice to estimate the hardness, roughness and cleavage strength of paper-based surfaces, the adhesion of coatings and even the surface soundness of wood-based products.

3.6.1 Bonding Tests

Most wood-based panels and structural timber elements are subject to clearly defined requirements regarding bonding quality. For many other wood and wood-based products applications, at least in Europe, bonding quality is largely unregulated—the only requirement being that the bond must be able to accommodate normal service loads in line with the intended purpose. It is ultimately left to the investigator, then, to decide what tests to apply in assessing, for example, the adhesive joints in garden furniture or a

FIGURE 3.49 Plywood that splits apart on its own after six hours' boiling needs no further tests to rule out standards-compliant water resistance for outdoor applications.

living room table, or the bondlines in a three-layer parquet product or between decorative veneer and substrate.

The investigation of bonds in components that have already been used should always take account of service conditions as a possible cause of overloading and damage. It must also be remembered that bonding failures often occur only locally. This means that, if all undamaged locations exhibit a strong bond, this may simply be because all the poorly bonded areas have already failed and are thus no longer available for adhesion testing.

A preliminary assessment of bond performance can be made as follows:

Adhesives normally exhibit a higher internal bond and cleavage strength than wood. Hence, if the bond is sound, the failure will tend to occur in the wood at some distance to the bondline rather than in the adhesive. The fracture behavior can be assessed by splitting open the joint in the grain direction, for example, using a firmer chisel.

A six-hour boil test will provide a general indication of whether the bonding in a wood-based product is water-resistant. If the test causes the wood-based product to split apart or exhibit extreme thickness swell, it may be assumed that the bond is not water-resistant (Fig. 3.49).

Although the chisel/knife tests described, with minor differences, for example, in BS 6566-8, ISO 12466-1 and AS/NZS 2098.2 allow only a general assessment of plywood bonding quality, experienced investigators can interpret the results with a reasonable degree of reliability (Fig. 3.50).

On the other hand, no further tests on bond performance are likely to be needed when analyzing products that are still intact after evidently long and intensive moisture exposure (Fig. 3.51).

3.6.2 Mechanical Tests on Coatings

It is important to remember, especially with regard to surface characteristics, that any specified requirements will always relate to the time of delivery. Often, however, faults

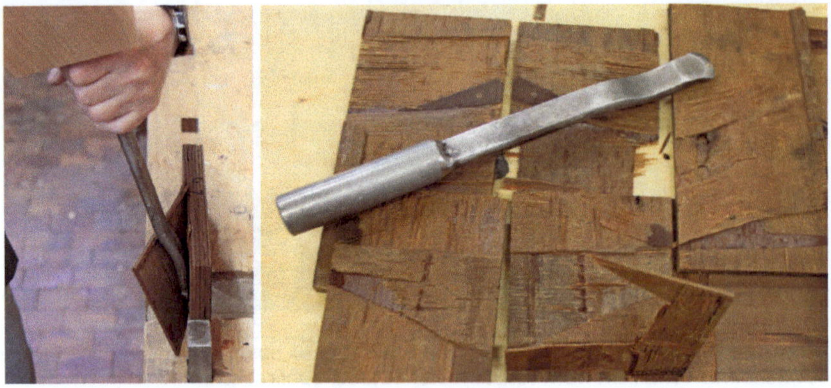

FIGURE 3.50 Plywood which, though remaining intact after six hours' boiling, is easily prized apart by a chisel along the bondlines, with hardly any wood fracture, offers a certain degree of moisture resistance. At the same time, it will certainly fail to meet the strict requirements for exterior-type (waterproof) plywood.

FIGURE 3.51 A particleboard panel severely damaged by a wood-destroying fungus (as indicated by the fruiting body) shows hardly any swelling at the edges. This is clear evidence of a very water-resistant adhesive.

are only found in surface finishes after use. Whether users are entitled to expect the same performance, for example, in terms of hardness or scratch resistance, as at the time of delivery—at least for the less exposed surfaces—must be judged on a case-by-case basis.

Adhesion and Wet Adhesion

Coatings applied to dry wood almost always achieve adequate adhesive strength. Problems typically occur where water permeates into the interface between coating and wood, with peeling, cracking or blistering as the consequence.

This makes wet adhesion (i.e., adhesion to wet substrates) one of the crucial properties for outdoor coatings. At the same time, the methods used to determine wet adhesion and, even more so, the interpretation of the associated results are far from unproblematic.

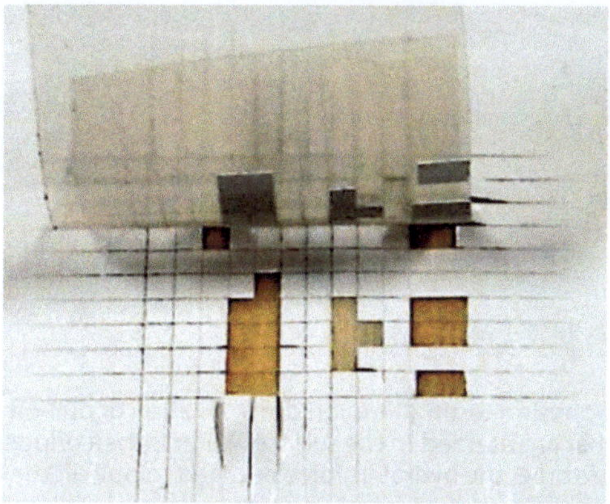

FIGURE 3.52 Cross-cut test: a sharp knife is used to make incisions through the coating. A specified adhesive tape is applied and pulled off, and the peel-off area assessed. For the wet adhesion test, the surface is moistened for two or so hours after incision, prior to applying and pulling off the adhesive tape.

In his book *Failure Analysis of Paints and Coatings*, Dwight G. Weldon delivers a clear verdict on adhesion tests:

There is often a great deal of confusion and controversy regarding adhesion testing. It can be argued that adhesion is the single most important property of any coating. Yet, after a century of modern paint technology, the measurement of adhesion is, at best, an imperfect science, subject to error and interpretation. [79]

While the cross-cut test (Fig. 3.52)—as described with minor differences in ISO 2409, ASTM D3359, AS 1580.408.4 and elsewhere—is easy enough to perform, the specified procedures are limited to dry adhesion tests. To assess wet adhesion, the sections are first moistened with a wet cloth for two or so hours before applying and pulling off the adhesive tape [58]. The cross-cut test assumes considerable experience and the results are not particularly reliable, even for tests conducted by expert investigators. For this reason, the method should only be used as a means of comparing different coatings, that is, to provide a general indication as to whether the coating under investigation adheres better or worse than the reference coating. The method is particularly suitable for identifying the zones of a surface with superior and inferior adhesion.

Exact numeric values are supplied by the adhesion test to ASTM D4541-02, which involves the use of portable pull-off adhesion testers (Fig. 3.53). The reproducibility of the results, however, is often poor and the significance of the results accordingly low. The following quotation by Dwight G. Weldon on the problematic nature of adhesion tests applies in equal measure to the tests under ISO, ASTM and AS/NZS standards:

Regardless of which method one prefers, there is still the problem of making sense of the results. The ASTM test methods are of little help here, for they report either psi or use an arbitrary rating system that runs from a "5" (the best result) to a "0" (the worst result). Is a "5" good and a "1" bad? Is 450 psi good and 350 psi bad? There are no simple answers to these questions. The real usefulness of adhesion testing is to see if there are any trends. [79]

FIGURE 3.53 Wet adhesion testing through determination of pull-off strength. Grooves are cut around so-called "dollies" that are bonded to the surface. Water is then dripped into the grooves. After a predefined exposure time, the hydraulic force required to pull off the dollies is measured.

A rough indication of coating adhesion can also be obtained when preparing thin sections with a microtome. Here, wet adhesion is unlikely to be particularly poor if the saturated coating continues to adhere to the wood during cutting (see Fig. 3.91).

Flexibility/Brittleness

In the context of coatings, the term "flexibility" is used to describe the ability of a dry film to deform elastically or plastically under mechanical loads, instead of suffering brittle fracture. This is an important property for coatings in that its counteracts stress whitening and cracking. The brittleness of practically all coatings increases over time. A key parameter for long-term serviceability is thus the flexibility after a certain period of use rather than flexibility shortly after application.

In most cases, the scarcity of codified criteria only allows coating flexibility to be assessed in comparison to other standard coating products or based on empirical values. Furthermore, given that it is not normally possible to remove coatings from the substrate without damage, all tests must also factor in the possible impact of the substrate on the results.

A preliminary indication of coating brittleness is very easy to obtain using the "coin test," whereby a coin is pressed firmly against the film and scraped across the surface (Fig. 3.54).

FIGURE 3.54 A coin is pressed firmly against the coating and scraped across the surface. The white color of the scuff mark (stress whitening) reveals that the coating is very brittle.

Here, stress whitening is a typical sign of poor (intercoat) adhesion or brittleness in the coating. As the test results are partly dictated by the coin shape, pressure and angle, a certain degree of familiarity with the method is needed for an accurate assessment of the coating properties. Yet, even inexperienced investigators can gain useful insights by comparing the results for different coatings.

An indication of coating flexibility can also be gained when preparing thin sections for microscopic analysis. Here, if the film remains intact instead of disintegrating when cut, it is unlikely to be particularly brittle (see Fig. 3.91).

More extensive analyses can be performed with special devices that are capable of determining the hardness of coatings, that is, their resistance to indentation or penetration by a solid object (microhardness).

Hailstone is a common cause of the premature failure of outdoor coatings. Even relatively small hailstones can penetrate coatings, leaving typically circular perforations (Fig. 3.57).

The damage caused by hailstone may be intensified by coating brittleness, poor adhesion to the substrate, and the degradation of lignin (a constituent of wood) by UV radiation, small coat thicknesses, and soft species of wood. Ball-impact and similar tests offer a useful means of determining the comparative sensitivity of different coatings to hailstone (Fig. 3.55). The aim of such tests is to establish the height from which a ball can be dropped onto a surface (or the force with which an indenter can hit a surface)

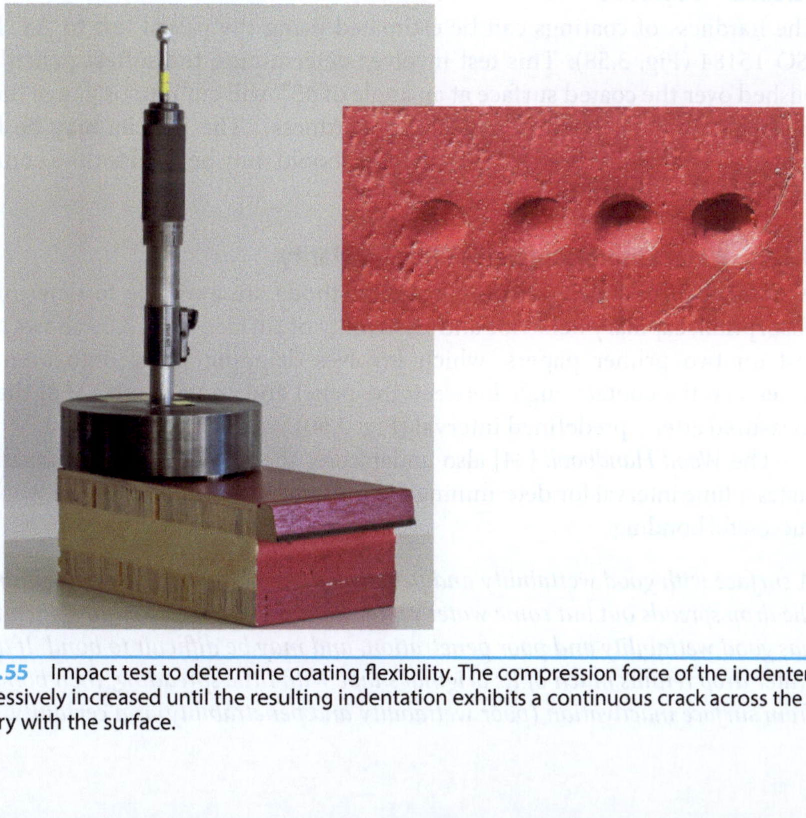

FIGURE 3.55 Impact test to determine coating flexibility. The compression force of the indenter spring is progressively increased until the resulting indentation exhibits a continuous crack across the boundary with the surface.

FIGURE 3.56 Comparative test on flexibility of two coatings.

without causing cracks. The test can also be applied to investigate the brittleness of indoor coatings by comparing the performance of different products.

Another comparative test of coating flexibility involves bending thin strips of material over various radii: the smaller the radius over which the strip can be bent without cracking, the less brittle the coating (Fig. 3.56).

Hardness—Pencil Test

The hardness of coatings can be estimated using the pencil test to ASTM D 3363 or ISO 15184 (Fig. 3.58). This test involves determining the softest pencil which, when pushed over the coated surface at an angle of 45°, will cut into or gouge the film (pencil hardness) or scratch the film (scratch hardness). The pencils may be hand-held or mounted in a special holder. This method should only be used for the comparative testing of different coatings.

3.6.3 Water Absorption/Wettability/Capillarity

A number of relatively uncomplicated methods are available to determine the water absorption, capillary behavior and wettability of surfaces. Fig. 3.59 shows a comparative test on two primer papers, which involves dripping water onto a surface. Where necessary, the contact angle between the panel and contact surface of the drop can be measured after a predefined interval (Fig. 3.60).

The *Wood Handbook* [34] also underscores the significance of this method and even states a time interval for determining the minimum wettability of solid wood surfaces for successful bonding:

A surface with good wettability and penetrability will absorb the drop within 20 s. If the drop spreads out but some water remains on the surface after 40 s, then the surface has good wettability and poor penetration, and may be difficult to bond. If after 40 s the water drop retains much of its original shape with little spreading, then bonding problems from surface inactivation (poor wettability and penetrability) is a certainty.

This statement should not readily be applied to materials other than solid wood. Experience suggests that water drops placed on wood-based panels containing hydrophobic (water-repellent) agents, such as wax, may not spread out for several minutes, though subsequently show no bonding problems.

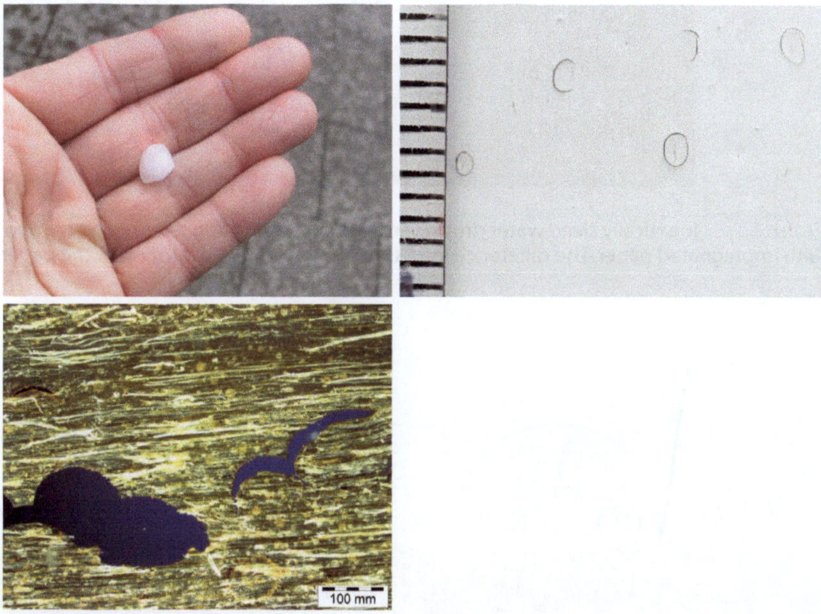

Figure 3.57 A hailstone and the typical near-circular marks made by impact on a white coating. The perforations caused by hailstones are clearly visible in the bottom image, which shows the underside of a detached semi-transparent coating. The adhering wood fiber also indicates that the coating failed to offer adequate UV resistance. Where coatings are brittle and substrates soft or pre-damaged, even pea-sized hailstones can bring about this type of failure.

Figure 3.58 Pencils with varying lead hardnesses (6H to 6B) are mounted in a special holder and pushed over the coated surface.

FIGURE 3.59 Identically sized water drops were simultaneously applied to two plywood boards faced with impregnated paper. The difference in wettability and water absorption of the papers is apparent.

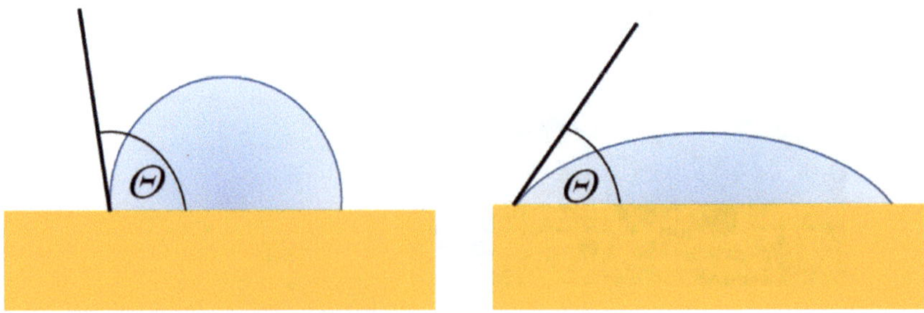

FIGURE 3.60 The contact angle between a liquid drop and the surface provides a useful indication of the surface characteristics: the greater the angle, the poorer the surface wettability. As this angle changes with time, especially on absorbent substrates such as wood, it is important to take the measurements after a predefined time interval. The angle can be estimated by photographing the drop from the side and analyzing the image with a suitable program.

The water absorption—and thus the condition—of low-build transparent coatings can be determined by observing the discoloration of wood when covered by a wet cloth: a darkening of the wood under the wet cloth after a few minutes signals high water absorption by the finish and the possible need for recoating.

Water absorption can be measured by weighing the specimens after predefined intervals. A stain can be added to the impregnating solution to establish the penetration depth (Figs. 3.61 and 3.62).

Dyes can be used to trace the path taken by water in the course of a failure. Here too, cotton blue stain has proved a useful means of coloring water, with fluorescent dyes as a further option. Figure 3.63 shows a demonstration, using a drop of stained water, of capillary effects on kitchen countertops. A large number of countertops displayed signs of swelling under the end caps of the wall trim. The end caps were fitted flush with the

Figure 3.61 Cotton blue stain was added to two colorless impregnations to record their penetration depth. After an hour's storage in the dyed solutions, the specimens were split apart lengthwise. The influence of the natural variability of wood was minimized by cutting all specimens from the same wood strip.

Figure 3.62 To record the water permeability of a coating, water stained with fluorescein was made to stand on the surface for several hours. Viewed under a fluorescence microscope, the above polished section shows the penetration of the stain.

ends of the countertops. The test showed how water was drawn into the capillaries between the end cap and the adhesive joint below this, which evidently caused the observed swelling.

Figure 3.64 pictures two microscope slides and some colored water, arranged to demonstrate the capillary effect.

3.6.4 Density

Density is the mass per unit volume of a substance, expressed in kg/m^3 or g/cm^3. Given that density is closely correlated to hardness and many strength properties of materials, it sometimes represents an important parameter in failure analyses.

A balance and a sliding caliper are all that is needed for a straightforward measurement of specimen density in accordance with many national and international standards. Depending on their condition, the specimens are prepared with a sharp knife, saw or plane—as far as possible with parallel-cut edges—and then precisely measured. Application of the method is independent of the water content of materials. At the same

Figure 3.63 A drop of blue-stained water is deposited at the joint between countertop and wall trim. After removal of the trim end cap, it can be seen how, by capillary action, the water is sucked directly onto the adhesive joint of the edge band. Due to the capillary effect, water is also drawn up between the end cap and wall trim.

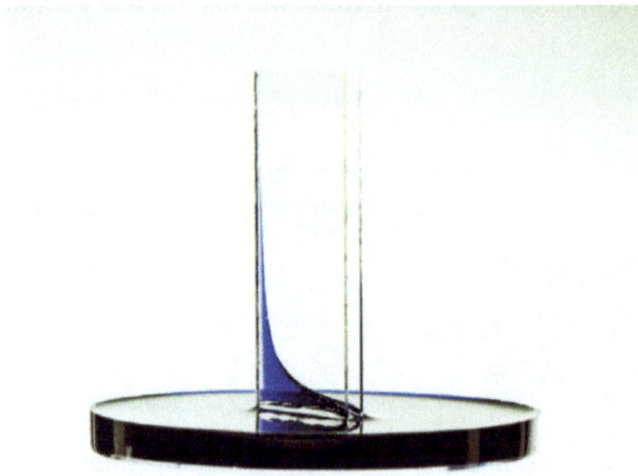

Figure 3.64 To demonstrate capillary action, two microscope slides were stood at an acute angle to each other in stained water. As the gap between the slides narrows, the water level rises in a hyperbolic trajectory. This illustrates the surprising height to which thin capillaries can draw up water.

time, the exact preparation and measurement of specimens can be very difficult where these are heavily decayed, have cracks or exhibit high slope of grain.

Wood density can also be estimated under the microscope using the systematic manual point method (see Sec. 3.1.8). This is because wood cell walls, regardless of species, have a uniform density of around 1.5 g/cm^2. The approximate density can thus be calculated by determining the ratio of cell walls to pore space.

The body weight of an adult, if applied as a point load, suffices to leave clear dents on virtually any wood species. Typical causes include defective shoe heels, studs on trousers (see Fig. 4.73) or grit on the floor. Here, a typical reaction on the part of building occupants is to cast doubt on the hardness of the wood. The Brinell hardness number is a good indicator of the likely depth of impressions made by falling objects or other mechanical loads in parquet surfaces [72]. For wood, however, Brinell hardness measurements are complex and the results relatively inaccurate, with wide deviations. Several authors, including [61], have demonstrated a close correlation between Brinell hardness and wood density. Density determination therefore offers an alternative and far more reliable method of estimating wood hardness.

Archimedes' Principle

Archimedes' principle can be used to establish the volume of irregular-shaped specimens [43]. Pieces of any size are attached to a thin needle and immersed in a vessel filled with water (Fig. 3.65). As water has a density of 1 g/cm^3, the mass increase in grams is equal to the volume of the specimen in cm^3. To minimize water absorption, dry specimens can be coated or treated with an impregnation spray prior to immersion.

Density Profiles of Wood-Based Products

Gamma-ray radiography offers one means of determining the density profile of wood-based products. The density profile allows conclusions to be drawn about the manufacturing conditions and properties of panels:

- Asymmetrical density profiles point to problems in the manufacturing process or suggest that one face of the finished panels was sanded to remove the press

FIGURE 3.65 Use of Archimedes' principle for volume determination. A specimen skewered on a needle displaces a quantity of water equal to its volume. The balance records a weight increase of 13.77 g, which indicates that the specimen volume is 13.77 cm^3. AS/NZS 2098.7 describes how to estimate density by comparing the length of the submerged portion of a test piece to the total length of the sample.

skin. Asymmetrical panels are susceptible to deformation and normally indicate shortcomings in production.

- A sharp drop in density in the core layer tends to lower the internal bond strength of panels. Inadequate internal bond strength can lead to splits and cracks in the edge surfaces.

A drop in the density profile at the outer edge of the surface layers normally indicates that the press skin was not properly sanded away (Fig. 3.67). This skin automatically forms during manufacture due to partial curing of the adhesive through direct contact with the hot press plates before the press is closed. The faster the press closes, the thinner the skin. As the surface of the press skin is hydrophobic (water-repellent), it cannot be satisfactorily bonded or coated. The skin also tends to be too loose for proper bonding and coating and needs to be sanded away to achieve an adequately strong, solid panel surface.

Except with certain deep-router grades of medium-density fiberboard (MDF), the density profile does not serve as a measure of product quality and only seldom allows the immediate identification of a failure cause. While abnormal density profiles may sometimes help to explain the specific behavior of a panel (Figs 3.66–3.68), they may equally well have absolutely no practical relevance and be found in panels that comply with all requirements of the relevant standards.

3.6.5 Estimation of Stresses

Compressive or tensile stresses in wood or wood-based products are normally invisible. The only external clues as to the stress conditions in the material are supplied by the deformations or fractures that serve to relieve these stresses (Fig. 3.69).

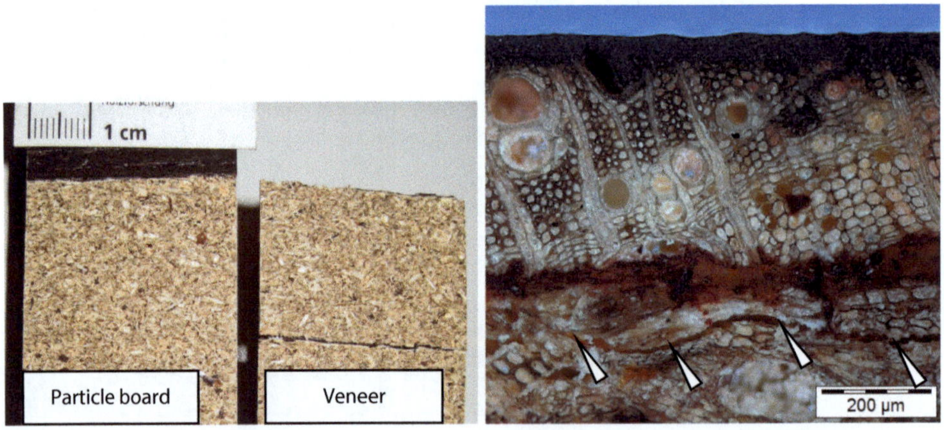

FIGURE 3.66 The veneer has peeled away from the particleboard substrate, with the surface of the particleboard stuck to its underside. The micrograph (right) shows the fracture zone within the particleboard. An investigation of the density profile is often useful where coatings or facings become detached from wood-based products and the fracture occurs within the wood-based product.

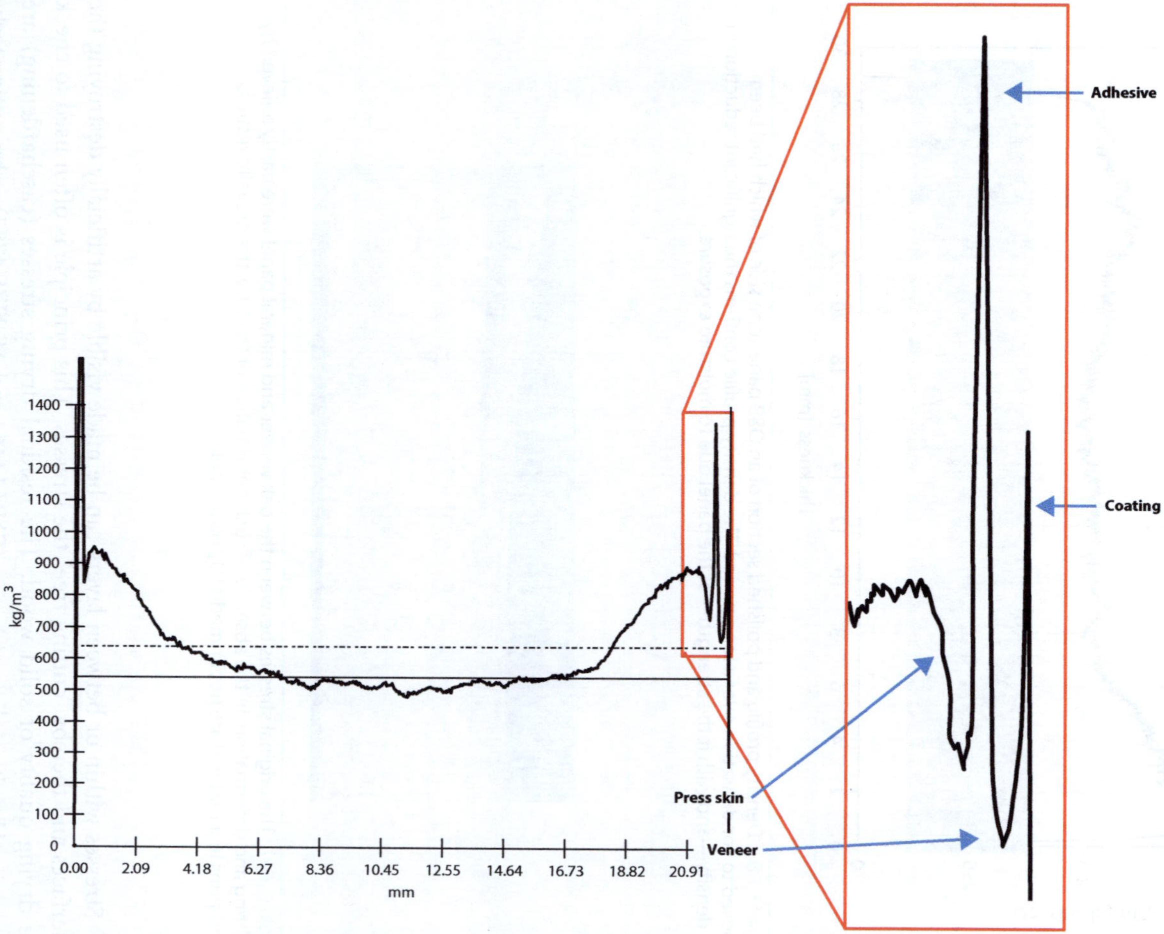

FIGURE 3.67 The density profile of the veneered particleboard in Fig. 3.66 shows a drop in density at the surface of the particleboard top layer. This clearly suggests that the press skin of the particleboard was not properly sanded away.

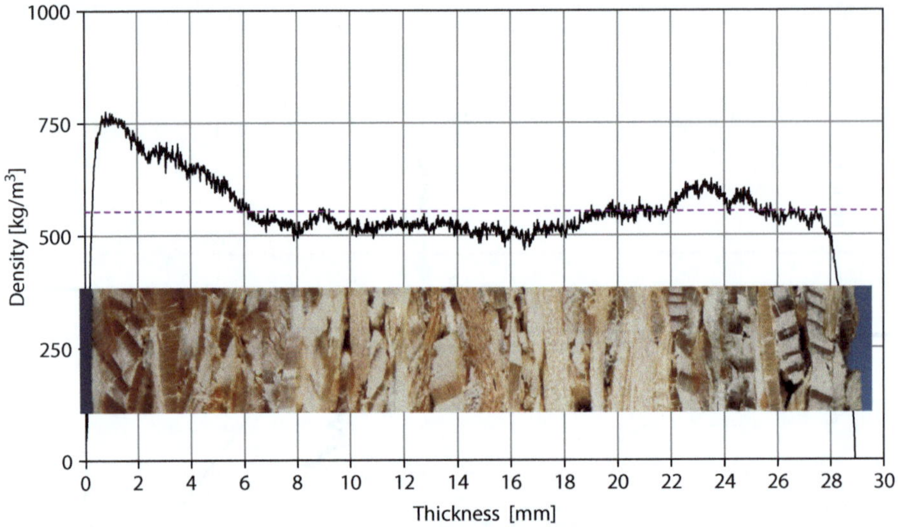

FIGURE 3.68 Density profile and polished section of an OSB panel, one side of which had been exposed to condensation over a long period. The density profile confirms the significant reduction in the density—and with it the strength—of the panel due to moisture exposure.

FIGURE 3.69 The original stresses between the oak veneer and mineral panel were partly relieved by splitting and deformation. In the absence of such deformation or fracture, the prevailing stress conditions in a material are not immediately observable.

Stresses within or between layers can be made visible by artificially destroying the interfaces and thereby "cutting free" the stresses. This principle is often used to check the drying quality of solid wood. The residual drying stresses (casehardening) are determined by either slicing/ripping (ENV 14464 and AS/NZS 4787) or by cutting fork-like samples [22].

The estimation of internal stresses by cutting free is often a useful aid in failure analysis, especially when investigating cracks and deformations in boards faced with

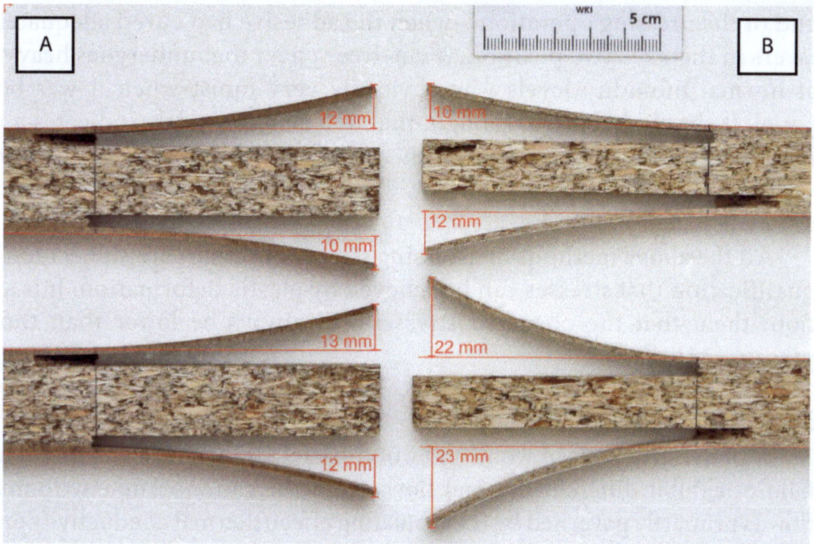

FIGURE 3.70 The cut-free surfaces of two panels faced with low-pressure laminate are pictured in the production direction of the laminate (top) and perpendicular to this (bottom) after 24 hours' storage at 70°C. The Panel B laminate is seen to exhibit greater deformation perpendicular to the production direction than the Panel A laminate. Only Panel B suffered cracking in service.

resin-impregnated papers and laminates. For all thin layers that cannot be separated from the substrate without damage, the cutting-free method is the only way to estimate stresses [65].

Resin-impregnated papers shrink when directly bonded to the substrate. The resulting tensile stresses may result in cracking if these stresses locally exceed the material strength. Given the high modulus of elasticity of resin-impregnated papers, even minor stress differences between the top and bottom sides of double-faced panels may lead to deformation. Where tensile stresses are present in the laminate, the cut-free layer will curl away from the panel. Large-scale deformation (compared to empirical values or the behavior of similar materials) indicates the presence of high stresses between the laminate and the surface layer of the substrate. The cutting-free method can also be used to assess the response of materials to changes in moisture. For example, cut-free specimens can be stored for 24 hours both under normal environmental conditions (e.g., 20°C/65% relative humidity) and at 70°C. Comparison with a representative number of similar materials, a different batch of the same material or even the opposite side of the same specimen may allow conclusions to be drawn about the product in question and its processing (Fig. 3.70).

Given the often clear correlation between moisture and stress, the cutting-free method also allows estimation of the moisture conditions prevailing during the manufacturing process, or more specifically the moisture at the exact moment—normally at the end of the pressing operation—when the adhesive had cured adequately to hold the layers in their relative position. A cut-free veneer that undergoes heavy deformation at normal humidity levels was obviously very moist when it was bonded to the panel. It should be noted, though, that deformation solely indicates a moisture difference, the most likely cause with veneers being a high level of moisture in the veneer itself. The deformation could, however, also be explained by an extremely dry base panel.

All the above methods of assessing stress conditions are subject to the important qualification that stresses can be relieved by plastic deformation. It is a fair assumption, then, that the observed stresses will always be lower than those originally present [35].

3.6.6 Thermal Flow Thermography

Many of the structures in wood and wood-based products that are useful for diagnosing failures exhibit different thermal flows from their immediate environment. Thermal flow is primarily governed by the insulating effect/thermal conductivity and heat capacity of the material in question. Any temperature variations occurring on surfaces can be made visible by thermographic cameras [60].

Given the strong insulating effect of air, thermal flows through zones of entrapped air significantly differ from those in surrounding areas without such inclusions. Where layers are split apart, a tiny cushion of air forms and the thermal contact is eliminated or reduced. The resulting differences in thermal behavior can be identified nondestructively using special thermographic techniques. Likewise detectable are zones with particularly high thermal conductivity (knots, moist areas, etc.).

A short thermal pulse is applied to the component under investigation using a radiant heater or flash lamp and the cooling behavior of the surface recorded on film. The pulse produces a thermal front that penetrates into the object while the surface cools down. If any imperfections with an insulating effect obstruct the flow of heat into the interior, the surface above the imperfection will remain warmer for a longer period (Fig. 3.71). The layer of air below a poorly bonded veneer, for example, will prevent rapid dissipation of the applied heat into the substrate. The imperfection is thus immediately detectable by a surface temperature only a few hundredths of a

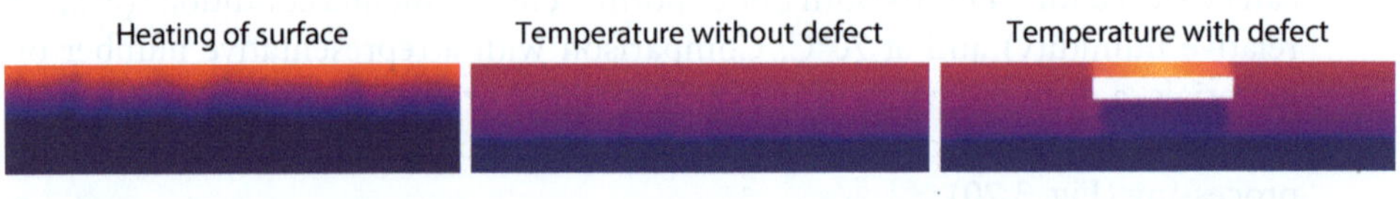

FIGURE 3.71 Temperature spread in a cross-section investigated using thermal flow thermography. The surface of the object is briefly heated up. As the thermal front penetrates into the object, the surface cools down. The position of an imperfection obstructing thermal flow into the interior is indicated by a warmer surface temperature.

degree higher than that of the surrounding area. Defects lying deeper below the surface will only show up in the thermographic image after a certain time. Depending on the thermal conductivity and heat capacity of the materials and variations in these parameters within the object, larger defects can be detected up to several centimeters below the surface [60].

The minute temperature differences encountered here are hardly measurable by the relatively inexpensive thermographic cameras normally used in construction. The above technique thus remains the preserve of specialists. One key application of thermal flow thermography is the investigation of valuable objects, for which nondestructive test methods are crucial. The procedure has also proved useful for preliminary investigations designed to identify those locations where more detailed analyses are likely to be informative. It also lends itself to applications where very large areas require examination. Figure 3.72 shows images from a thermographic survey performed on large wall areas. A decision had to be taken as to whether the visible imperfections simply needed a touch-up or whether full-scale renovation was required.

Figure 3.73 shows images from the thermographic analysis of a glider wing. The technique clearly reveals the quality of the bond between the aircraft plywood paneling and the support frame.

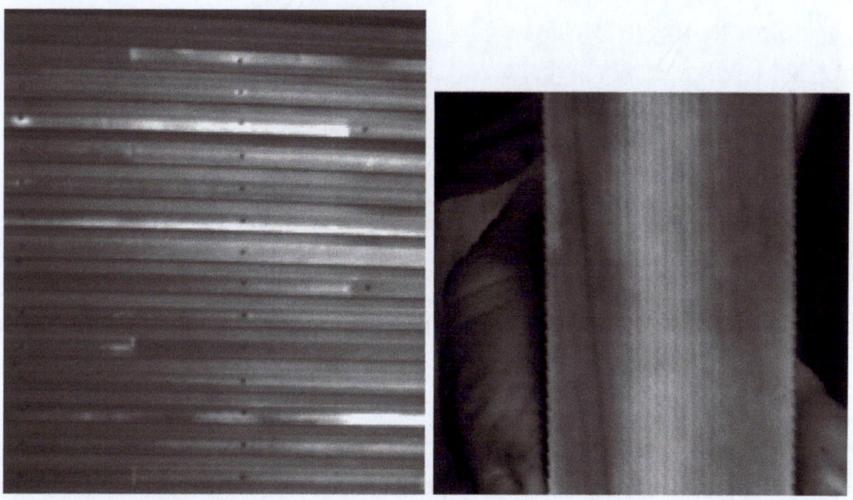

FIGURE 3.72 Thermographic image of a wall area comprising veneered wood strips. Various points in the wall and ceiling finishes, which spanned several hundred square meters, exhibited failures where the veneer had peeled away from the strips. During the site visit, heat was applied to some of the walls using a radiant heater. The light-colored areas in the thermographic images indicate those zones that are not bonded to the substrate. Many of these locations would not have been identifiable by visual inspection. The close-up thermographic image on the right reveals the failure cause: uncompressed adhesive beads along the middle of the strip.

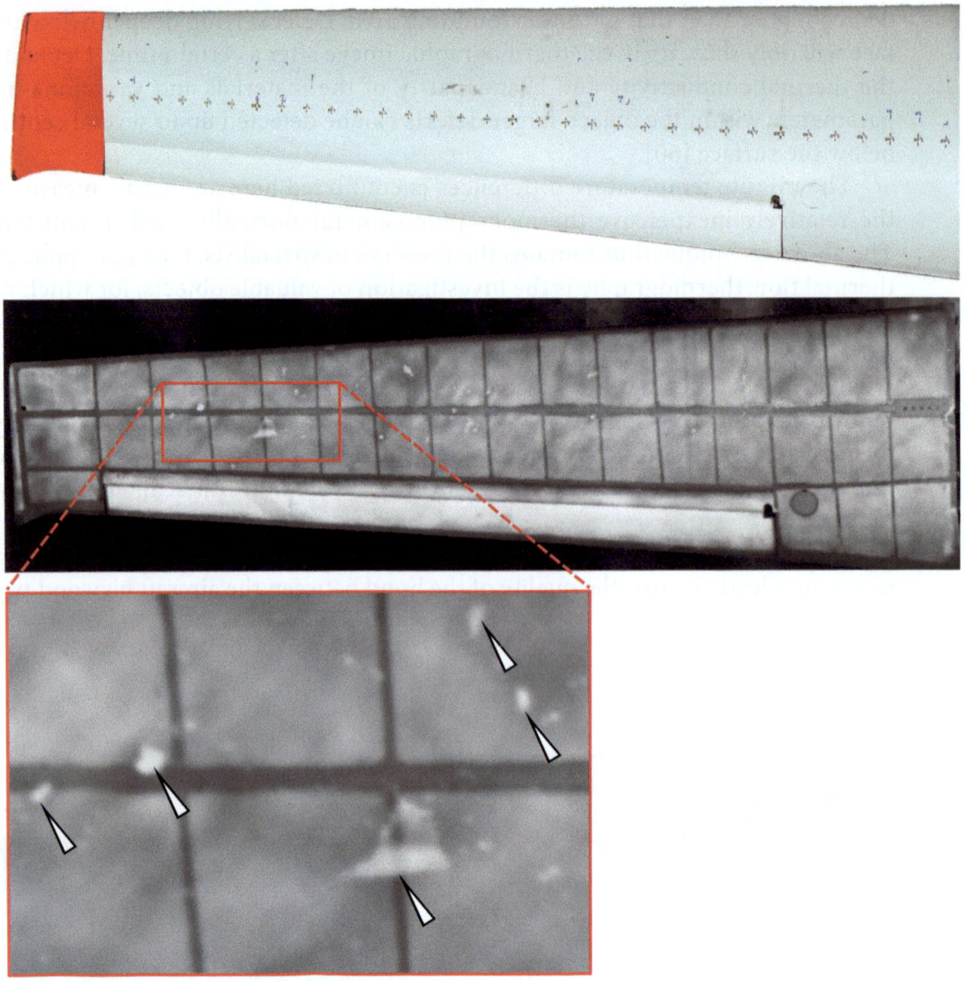

FIGURE 3.73 Thermal flow thermography used to analyze 3.7-meter-long wing of a glider. The small unbonded areas in the plywood paneling are recognizable by their light color. The layout of the support ribs and their strong bond with the plywood panels are indicated by the dark-colored pattern.

3.7 Microscopy

Microscopic examinations are indispensable for the clarification and documentation of many failure causes. Some issues can be resolved using only very simple microscopes, without expensive special equipment. Armed with four sheets of sandpaper (grit size P220, P500, P1000, and P3000) and a USB microscope possibly costing less than USD 100, a committed investigator can carry out analyses which, at one time, could only have been entrusted to professional laboratories.

This is not to say that specialized microscope techniques have become any less important. In particular, polarized-light microscopy, fluorescence microscopy and

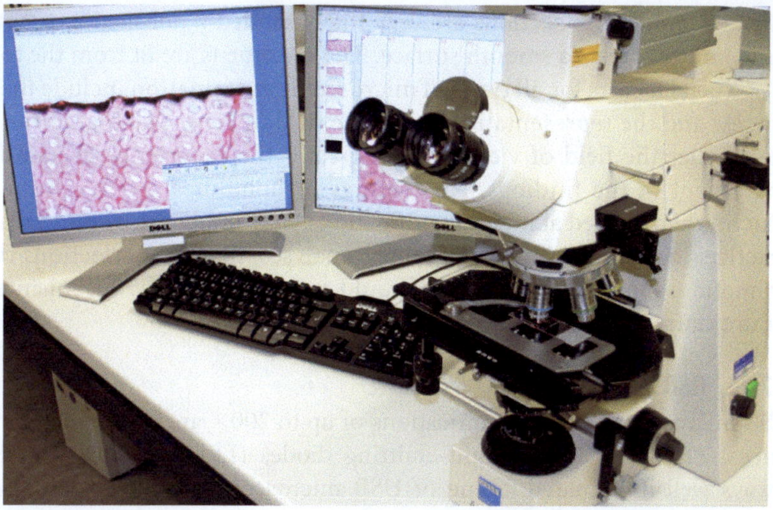

FIGURE 3.74 Universal microscope for transmitted-light and reflected-light microscopy.

certain staining methods sometimes offer the only reliable means of resolving questions. More elaborate techniques, such as phase contrast, are seldom essential and are not discussed in further detail here.

In rare cases, examinations under a scanning electron microscope (SEM) may also help to clarify the causes of failures in wood and wood-based products. Although SEMs offer magnifications up to 100 times greater than optical microscopes, their use is normally confined to the determination of chemical elements present at a certain location, based on EDX techniques (see Sec. 3.8.3). SEMs have the following disadvantages compared to optical microscopes:

- The images are not in color. Some layers or particles that are immediately discernible under an optical microscope on account of their color remain invisible under the SEM.
- Standard SEMs require a very high vacuum, which will extract all moisture from the specimens. Given the sensitivity of wood and wood-based products, it is always possible that certain cracks or other phenomena are the direct result of this drying process.

For this reason, electron microscope methods are omitted from the following discussion.

With regard to the optical microscope techniques presented here, a fundamental distinction must first be drawn between transmitted-light and reflected-light microscopy (Fig. 3.74). Examination under a transmitted-light microscope requires the preparation of thin sections from the specimens. These can be cut with a special device called a microtome or manually, by skilled investigators, using a razor blade. Reflected-light techniques, on the other hand, also allow the examination of unprepared surfaces. As a rule, however, cross-sections are very finely sanded ("polished") or cut with a microtome knife to create a smooth surface. The specimens are lit from the top or the side.

Crucial factors for all types of microscopic examination include the sample-taking process and the representativeness of the field of view. The higher the magnification, the smaller the field of view and the greater the experience needed for an accurate assessment of the findings. No coatings or bondlines are free from imperfections, which, when viewed at high magnifications, may imply problems that are irrelevant in practice. In the performance of failure analyses, then, imperfections that may appear dramatic under the microscope should not automatically be misinterpreted as the failure cause.

3.7.1 USB Microscopes

USB microscopes with magnifications of up to 200× are obtainable at surprisingly low prices. The incorporated light-emitting diodes (LEDs) ensure that the specimen is always well illuminated. Some of USB microscopes include polarizing filters to cut down reflections while other (albeit more expensive) models work with normal light and UV illumination for fluorescence excitation. Even basic USB microscopes can deliver impressive findings, such as coat thicknesses measured to an accuracy of around ±5 μm. These microscopes are additionally suitable for inspection and documentation purposes during site visits (Fig. 3.75). They also, to some extent, offer a useful alternative to endoscopes given that they are able to produce normal, unmagnified images and, thanks to their compact design, can be used at otherwise inaccessible locations. Some of the cameras allow direct calibration by means of scales to facilitate the exact measurement of images on the computer screen. For cameras without this feature, actual sizes have to be estimated by arithmetic means.

FIGURE 3.75 Use of USB microscope during site visit and in laboratory.

The following considerations can also help to estimate the sizes on USB micrographs:

- Softwood cells are approximately 30 μm wide in the tangential direction (Fig. 3.76).
- Although the size of human head hairs varies from person to person, the value is relatively constant for any particular head. Hairs from one's own head can thus be used as an indicative scale (Fig. 3.77).

Figures 3.78 and 3.79 illustrate how USB microscopes are sometimes able to deliver images only slightly inferior to those achievable using standard reflected-light microscopes.

Figure 3.76 Micrograph taken with a 1.3-megapixel USB microscope of a coating on a softwood cross-section polished to grit 3000. Softwood cells, which have a uniform tangential width of around 30 μm, can be used as an indicative scale. The thickness of the gray priming coat is thus estimated at around 130 μm and that of the white finishing coat at approximately 40 μm.

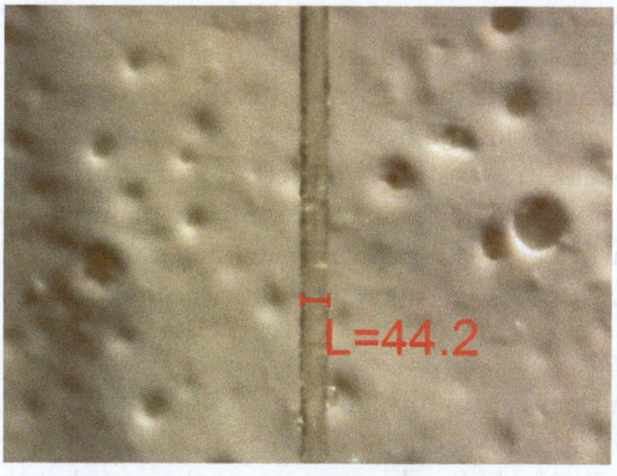

Figure 3.77 An approximately 45-μm-thick hair used to estimate sizes on a painted surface. The diameter of human hairs ranges between 40 μm and 120 μm. There is, unfortunately, no way of suppressing the decimal place shown by the USB camera, even though it suggests a nonexistent accuracy.

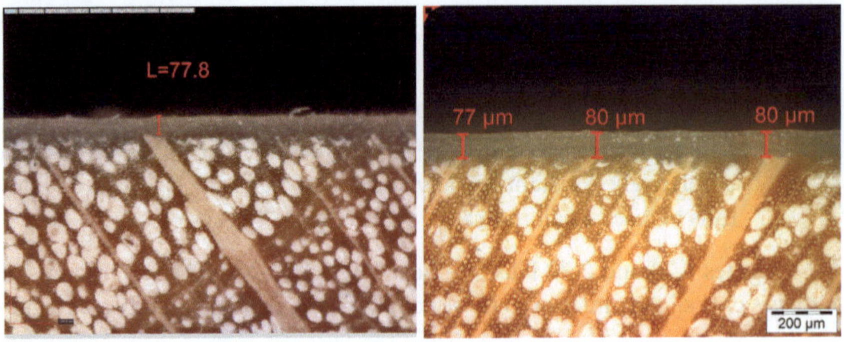

Figure 3.78 Polished section of coated beech parquet. Left: Micrograph from a 1.3-megapixel USB microscope. Right: Micrograph of same specimen from a reflected-light microscope.

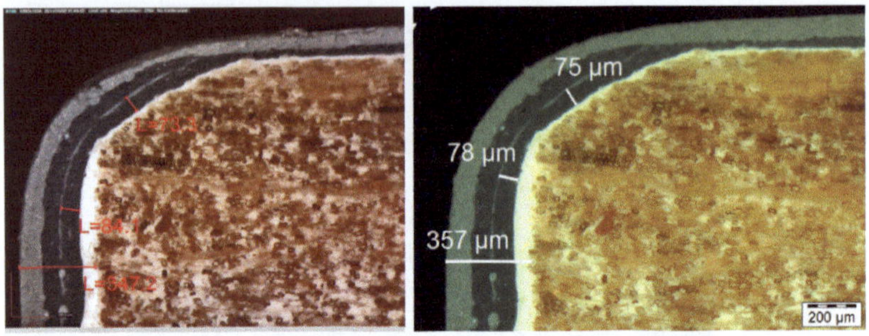

Figure 3.79 Polished section of coated MDF batten. Left: Micrograph from a 1.3-megapixel USB microscope. Right: Micrograph of same field of view from a reflected-light microscope.

3.7.2 Specimen Preparation

Traditional wood microscopy focuses almost exclusively on the analysis of thin sections, with any necessary examination of solid specimens being performed under an electron microscope. However useful this approach may be for advanced scientific research without financial or time pressure, it is frequently unsuitable for failure analysis where the focus is on gaining an overview, establishing the distribution of features and achieving a reasonable trade-off between effort and results. Moreover, the production of thin sections is not possible for many composite materials and wood-based products (e.g., HPL on particleboard). Practically any material, however, will allow the preparation of polished surfaces for fast and straightforward microscopic examination.

Polished Surfaces

Polished surfaces and polished thin sections offer a useful solution for the investigation of larger cross-sections or particularly hard layers/coatings and problematic substrates. Particularly advantageous is the fact that the specimens remain dry and are normally subjected to lower mechanical loads by polishing than by cutting. Polished surface specimens can be considerably larger than typical thin sections.

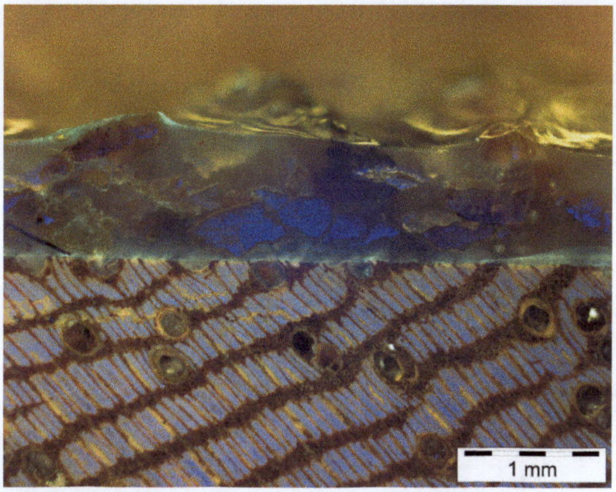

Figure 3.80 Polished section of a bongossi plank from a bridge (UV fluorescence). The coating comprises epoxy resin sprinkled with sand. Materials such as this cannot be cut with normal microtome knives.

Preparing a polished surface over the full cross-section of a window scantling takes around 5 minutes. The scantling cross-section can then be examined in its entirety to establish the composition and thickness of its various layers. Window sections of this type have a circumference of almost 20 cm (see Fig. 3.6). If the maximum width of microtome sections is assumed to be 10 mm, some 20 prepared sections would be needed to cover the same area. One downside of the method is the limited magnification achievable by reflected-light microscopes. Here, magnification factors are generally limited to 500 while transmitted-light microscopes still produce useful results at around 1,200-fold magnifications. Moreover, thin sections are indispensable for some staining methods, genuine polarized-light microscopy and other microscope techniques. Under normal illumination, transmitted-light microscopes also tend to deliver much more brilliant images (see Fig. 3.87). Yet, especially in conjunction with fluorescent methods, polished reflected-light specimens are able to achieve a similar color intensity and sharpness. Reflected-light microscopy may thus be regarded as an uncomplicated standard technology that is convenient for most routine applications (Fig. 3.80).

As a rule, specimens roughly 6–20 mm wide, 20–80 mm long, and 15 mm high are sanded for around 30 seconds at a time with papers of increasing fineness, for example, P220, P500, P1000, and P3000 grit, (Fig. 3.81). Before advancing to the next grit, the specimens are blown down as required to remove the sanding dust. It is important to avoid tilting the specimen. This is possible with a little practice provided the specimens are not too small. Extremely short or narrow specimens require a good deal of experience. Many sandpapers will leave grit particles on the specimen, which impair the quality of micrographs and may conceal findings. As experience has shown, the only way round this problem is to use sandpapers selected from different manufacturers as no single manufacturer appears to supply suitable papers in all grits.

FIGURE 3.81 Polishing by hand. Suitable papers include P220, P500, P1000, and P3000 grits. It is best to sand in the longitudinal direction of the specimen as this minimizes the risk of tilting. To remove the sanding marks of the previous grit, however, the specimen can also be sanded in the perpendicular direction or in a circular motion. The base should be absolutely even and only slight pressure should be applied to the specimen. Practice makes perfect!

The white sanding dust produced with this method gets caught in all surface voids and, if at all, can only be removed by thorough blowing down with compressed air. The white color of the dust makes even the smallest hollows and blisters stand out on polished areas (Fig. 3.82). Where the surface needs to be kept free of sanding dust, it can be pretreated with a suitable resin, for example, cyanoacrylate, epoxy or polyester, to fill the voids.

Very small specimens can first be cast in synthetic resin. Alternatively, they can be glued to a base block, for example, using cyanoacrylate adhesive, for easier handling during the polishing operation (Fig. 3.83). Given the small depth of field of microscopes, it is essential that the surface under investigation lies perfectly flat in the image plane. This can be achieved by pressing the specimens into a suitable soft compound with a sample leveling press (Fig. 3.84).

Should several thin specimens need polishing, these can be glued together into a single block. This has the advantage of allowing simultaneous microscopic examination of several specimens, making it easier to pick out differences between them. This procedure also ensures that the mechanical loads imposed during preparation are identical for all specimens. Potential sources of misinterpretation, such as cell deformation during the preparation of thin sections, are thereby eliminated (Figs. 3.85 and 3.86).

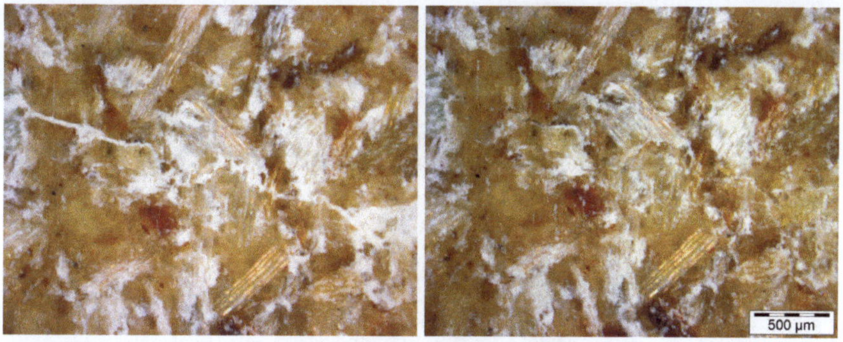

FIGURE 3.82 Polished particleboard surface. Left: The crack in the panel surface is easily discernible prior to blowing down with compressed air. Right: Without the contrast provided by sanding dust, the crack becomes almost invisible.

FIGURE 3.83 To facilitate polishing, very small specimens can be glued to a wooden block.

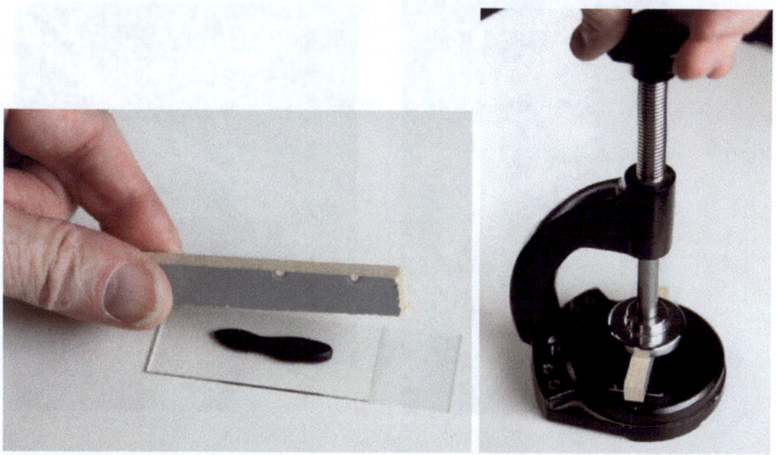

FIGURE 3.84 A polished section is mounted on a microscope slide using a suitable compound. A sample-leveling press ensures that the polished surface is exactly parallel to the microscope slide.

FIGURE 3.85 To facilitate polishing, several veneers made from the wood species wenge are glued together into a block.

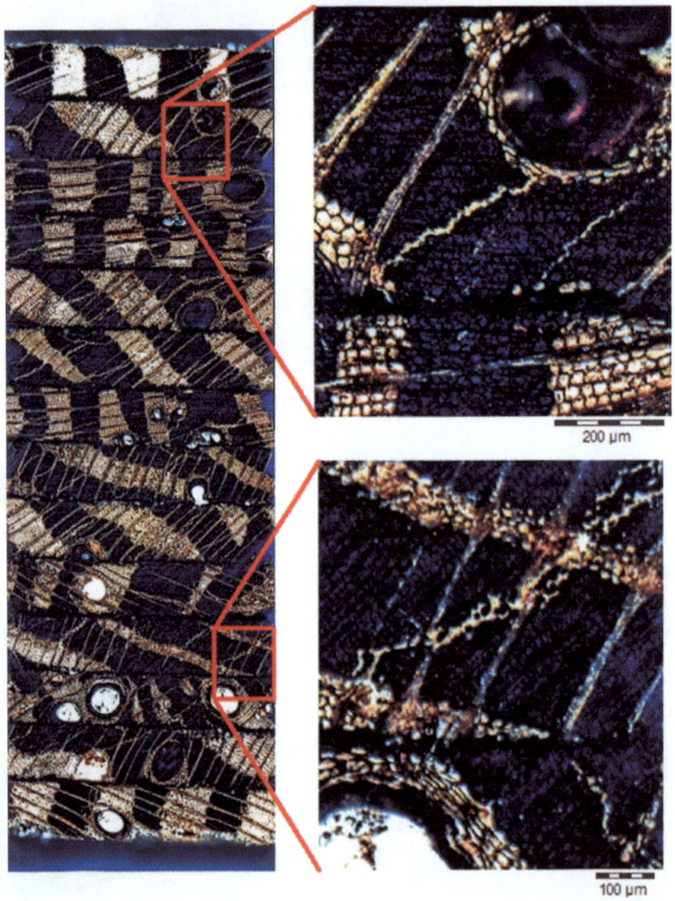

FIGURE 3.86 To compare the checks in several veneers made from the wood species wenge, these were glued together into a block, polished and examined under a reflected-light microscope. The micrographs were edited to enhance the contrasts.

Microfoam and other small voids in soft coatings and soft glues are sometimes almost invisible after polishing. If such voids are of special interest, the specimens are best prepared by cutting with a microtome knife or by hand using a razor blade. These sections may then be examined under a transmitted-light microscope. The results provided by thin sections are far superior in such cases (Fig. 3.87).

Polished Thin Sections

It is also possible to commission special companies to prepare polished thin sections from specimens that cannot otherwise be cut (Fig. 3.88). Yet, as these companies mainly focus on producing polished thin sections of minerals and the deployed embedding and grinding methods are accordingly geared to this type of material, not all providers deliver satisfactory results for wood specimens. Such sections are normally around

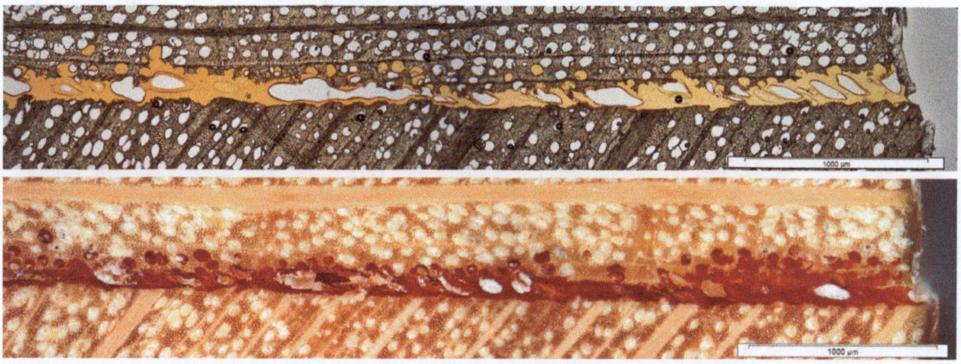

FIGURE 3.87 Transmitted-light micrograph of a thin section (top) and reflected-light micrograph of a polished specimen (bottom) showing same PVAc bondline with large voids. In both images, the PVAc is stained with Lugol's iodine.

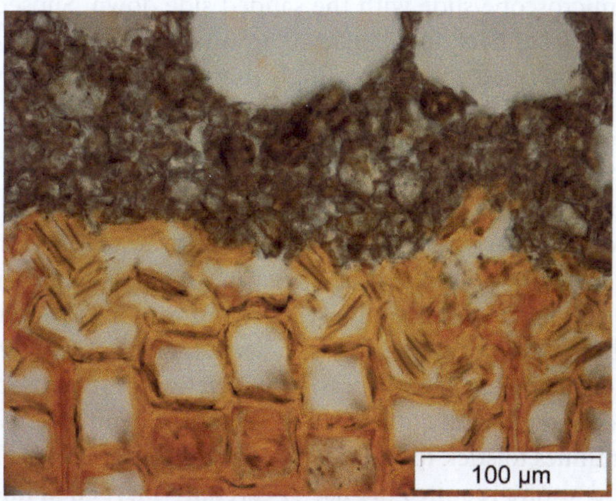

FIGURE 3.88 Polished thin section of cement layer on Douglas fir plywood.

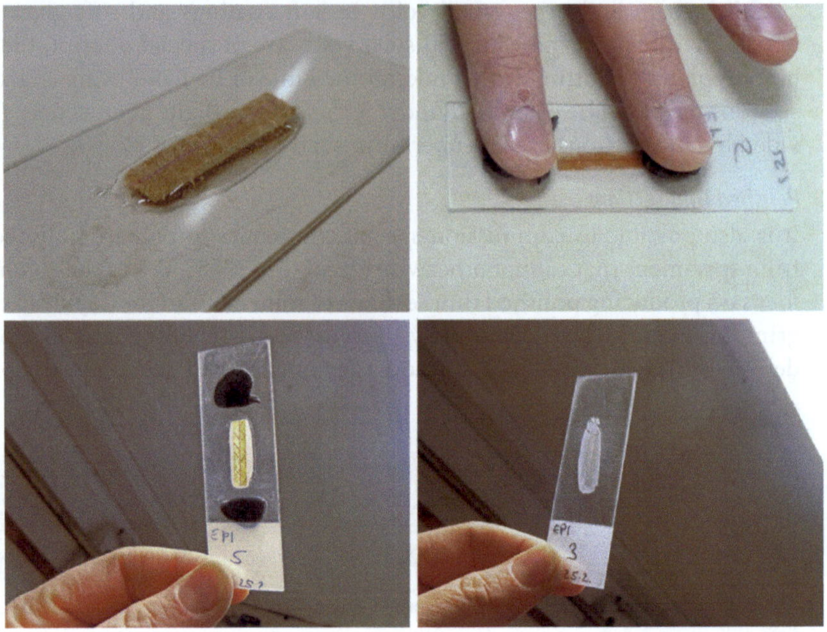

FIGURE 3.89 Preparation of a polished thin section by hand. The specimen is bonded, for example, with epoxy resin, to a microscope slide. It is first sanded, for example, with 220-grit, then with 1000- and 3000-grit papers until it becomes almost invisible. Two small blobs of plasticine facilitate handling of the slide during sanding.

20 μm thick. It is sometimes appropriate to use a colored embedding resin to provide a clearer contrast with the original specimen.

With a little practice, polished thin sections can also be prepared by hand. The surface under investigation is sanded to grit P3000 and bonded, for example, with epoxy resin, to a microscope slide with the sanded side down. Subsequently (or in advance), the specimen is cut back to a height of around 1 mm above the microscope slide. It is then carefully polished with increasingly fine-grit papers until, with the finest grit, only a wafer-thin layer of the section remains (Fig. 3.89).

Thin Sections

Thin sections can be prepared using a special cutting device called a microtome or also by hand. Hoadley [40] describes how thin pieces of wood can be manually cut using razor blades or very sharp knives in order to establish the wood species. The same technique can also be adopted to prepare thin sections, for example, of painted wood (Fig. 3.90).

Even the process of preparing thin sections can provide valuable clues about the properties of the finish. The successful production of intact sections, without any flaking of the coating during cutting, suggests that it cannot be particularly brittle. If the coating continues to stick to the face veneer, then adhesion is also adequate (Fig. 3.91). This method always requires the use of carefully sharpened microtome knives or fresh razor blades plus a good deal of experience in assessing the results.

Figure 3.90 Preparation of a thin section by hand. The zone to be cut is first dampened and then incised at a very shallow angle. The micrographs show the small flake lying on the razor blade. One advantage of cutting by hand is the automatic creation of extremely thin layers at the section ends: as the example shows, the internal structure of the coating is easily recognizable at this location.

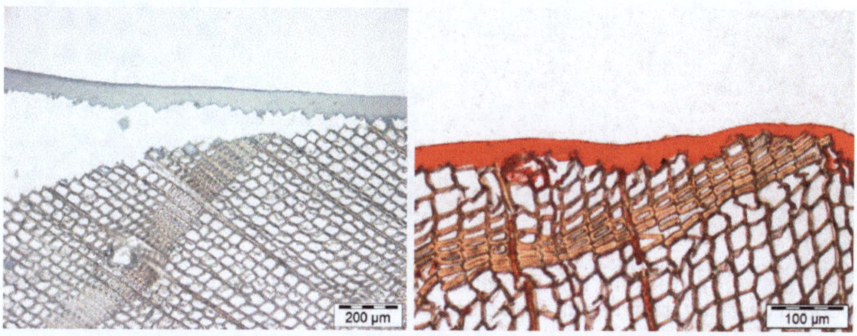

Figure 3.91 Left: Part of the paint film peeled away from the substrate during cutting, though without flaking. At least initially, it is reasonable to assume that the coating is not brittle, but with poor to, at best, satisfactory wet adhesion. Right: Non-brittle film with evidently good wet adhesion.

Thin sections are indispensable, among other things, for the microscopic identification of wood species, the assessment of fungal or bacterial attack and the detailed investigation of coatings (Figs. 3.92 and 3.93). Sensitive specimens can first be embedded in polyethylene glycol (PEG 2000) [71].

3.7.3 Illumination Techniques

Suitable illumination techniques can help to highlight the structures under investigation through the intensification of contrasts. Indeed, certain features are only visible under special lighting conditions, as with polarized-light or fluorescence microscopy. As a rule, trial and error is the only way of determining the best method in any particular case. Occasionally, lighting combinations that initially appear nonsensical prove to

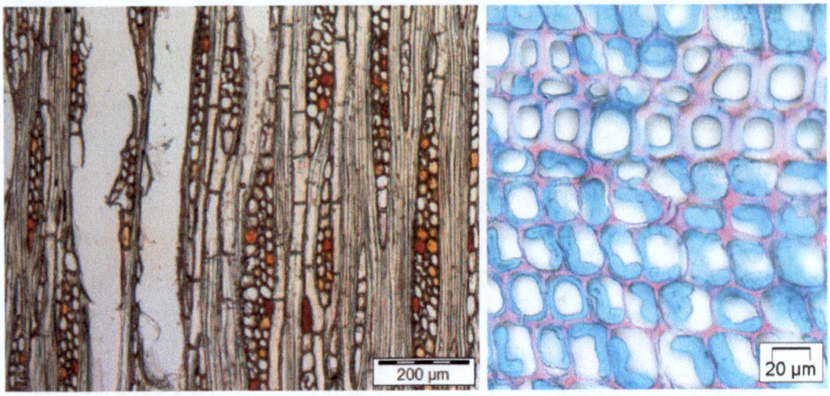

FIGURE 3.92 Thin sections. Left: For determination of wood species (abura). Right: Bacterial attack in pine (safranin/astra blue stain).

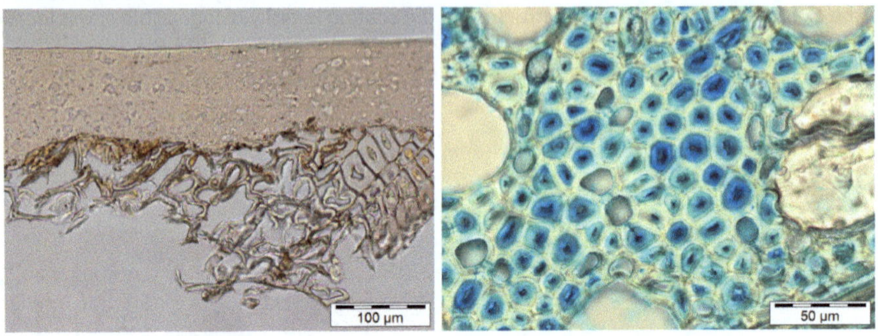

FIGURE 3.93 Thin sections. Left: For detailed investigation of coatings. Right: Identification of tension wood in beech by staining with astra blue.

be the best means of recognizing particular structures. Also, normally unwanted phenomena such as blur, color streaks, reflections or air bubbles can sometimes actually contribute to the information content of an image. While some techniques necessitate the use of elaborate equipment, many of the methods are relatively uncomplicated.

Among the many possibilities, the following illumination techniques have proved particularly useful for the failure analysis of wood and wood-based products:

- Normal light (bright-field and dark-field)
- Polarized light
- Fluorescence

The precise way in which these illumination methods work is discussed in detail in the relevant literature. One lesser-known, though often very helpful technique involves the use of an extreme form of bright-field illumination to produce pictures, which—for want of a better term—will be referred to here as "glint images." Glint images result

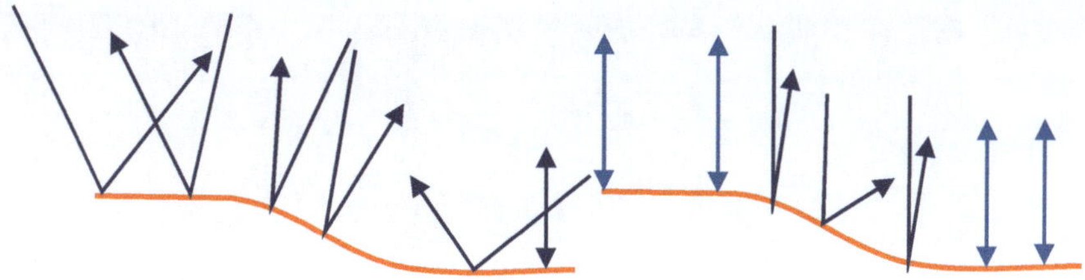

FIGURE 3.94 Under normal diffuse lighting conditions (left), brightness is largely independent of the surface structure. With directional illumination (right), even the slightest angle deviations result in strong brightness contrasts that highlight irregularities in the surface structure.

where the viewing angle is identical to the angle at which a directional light beam hits a glossy surface. The effect is independent of the scale of the object. It is witnessed in everyday life, for example, where light from a fluorescent tube reflects off a painted finish to reveal even the finest surface irregularities. The phenomenon is also familiar in the context of satellite technology under the name "sunglint." If sunlight hits a water surface at exactly the same angle as the viewing angle of a satellite opposite the sun, the observed area will display features that cannot be picked up by other sensors (http://www.nasa.gov/multimedia/imagegallery/image_feature_484.html).

Glint micrographs result when specimens are illuminated directly from above. This arises, in particular, where the light from a reflected-light microscope is projected onto the specimen through the lens. The resulting reflections reveal extremely fine details on the specimen surface that would otherwise only be discernible using an electron microscope or similar equipment. This illumination technique, however, entails the loss of color information. The contrast is produced by extreme differences in brightness resulting from even the slightest angle deviations on the surface (Fig. 3.94).

Under glint illumination, the proportion of light reflected directly from the surface of transparent coatings is so great that it completely outshines the light passing through the coating and reflected by the underlying wood surface. As a result, particularly where reflected-light microscopy is used to analyze transparent films, the two illumination techniques—normal and glint—often deliver excellent complementary micrographs of a completely different nature. While glint illumination delivers a detailed image of the coating topography in gray or blue shades, the normal image shows the surface of the wood in color (Fig. 3.95).

As glint is reduced by cross-polarization, some reflected-light microscopes allow the effect to be "switched" on and off simply by rotating the polarizer.

As normal stereo microscopes and macroscopes use side illumination, they are unable to supply glint images. However, by positioning a small LED in front of the lens, it is possible to create the lighting conditions needed for these images. Figure 3.96 shows surface images of transparent coatings taken under normal lighting conditions and the glint micrograph produced with the help of a small LED.

Combined Reflected/Transmitted Light

Some coatings remain opaque even in the form of thin sections and thus appear black under a transmitted-light microscope. To record such specimens with the natural color

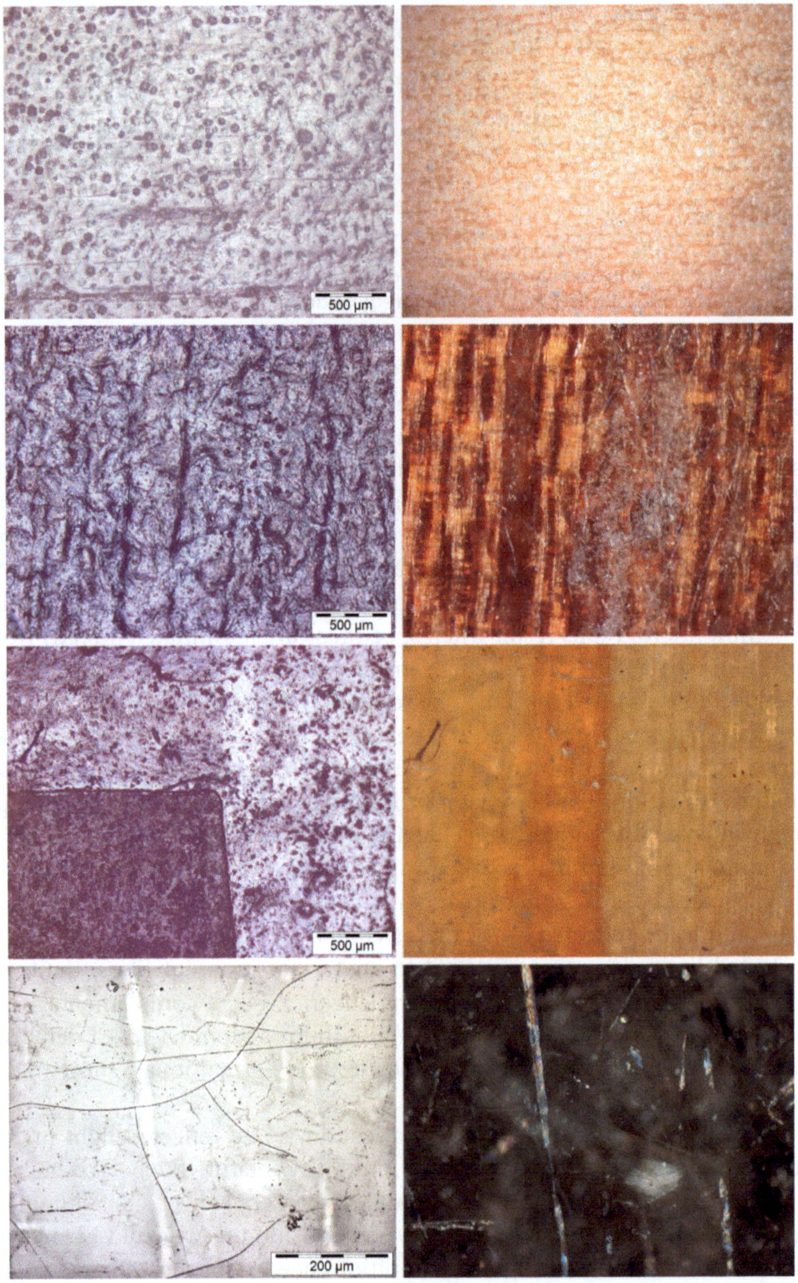

FIGURE 3.95 The left-hand micrographs show a transparent coating or, in the bottom example, a high-gloss CPL finish under glint illumination. The same surfaces are pictured in the normal images on the right. This underscores the often vast differences between the micrographs and information obtainable using the two techniques.

of the coating, it is possible to illuminate them simultaneously from the top and bottom. The translucent wood is captured by the transmitted light and the opaque coating by the reflected light (Fig. 3.97).

Fluorescence Microscopy

Fluorescent materials absorb photons of one wavelength, then reemit this energy at a longer wavelength. Fluorescence microscopy involves the use of special filters to block

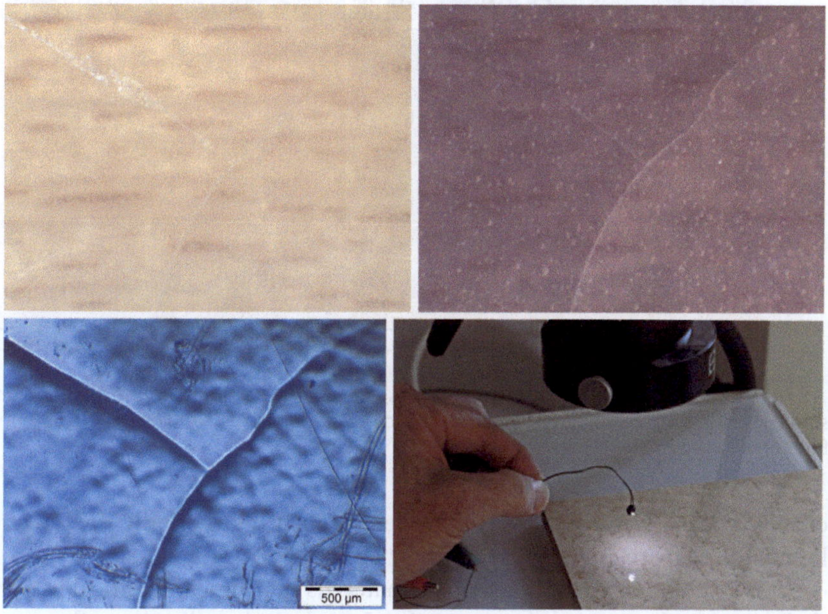

Figure 3.96 Surface of transparent coating viewed under a stereo microscope. Top left: Ring light illumination. Top right: Extreme oblique illumination. Bottom left: Glint micrograph produced by illumination with a small LED (bottom right).

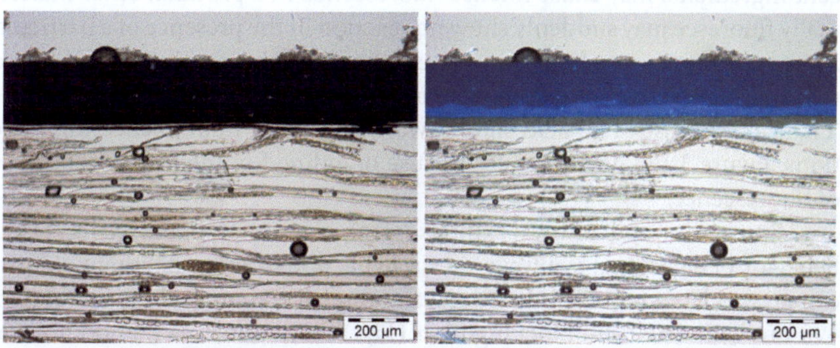

Figure 3.97 Left: Transmitted-light micrograph of thin section of coated wood. Right: A combination of the transmitted- and reflected-light techniques brings out both the wood features and the coating structure. The round shapes are air bubbles entrapped during preparation.

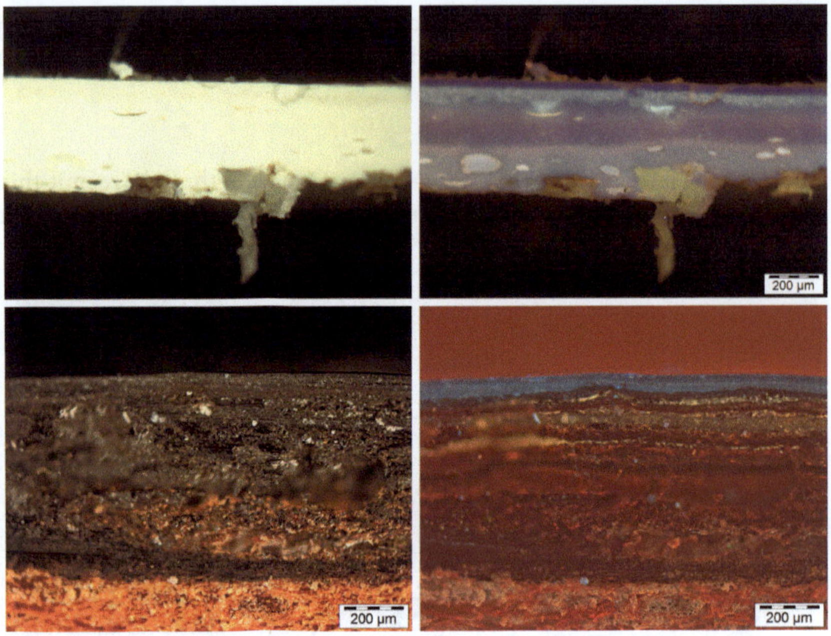

FIGURE 3.98 Under normal lighting conditions, the coatings pictured on the left appear to consist of a single layer. In both cases, however, fluorescence excitation at 450–490 nm with a 520-nm blocking filter reveals the existence of several layers.

out light at the original wavelength, thereby ensuring that only the longer-wavelength light emitted by the object reaches the viewer's eye.

It is often possible to differentiate materials that appear identical in normal light by their varying fluorescent characteristics. In practical applications, fluorescence is used to distinguish materials through contrast. At the same time, it rarely lends itself to direct diagnosis as it is impossible to predict what materials in standard commercial products will exhibit fluorescence at what excitation wavelength. While very low traces of fluorescent ingredients may cause intense fluorescence in a product, other materials that normally fluoresce may suddenly show no reaction in the presence of a particular substance. Moreover, fluorescent behavior sometimes changes over time. With the "dirty" specimens typically encountered in wood analysis practice, then, the focus is on interpreting "chance" contrasts. Fig. 3.98 shows two examples of fluorescence microscopy applications. Figure 3.133 demonstrates the use of iodine fuming for the selective suppression of fluorescence.

Untreated wood always exhibits some degree of fluorescence. The fluorescent behavior of algae, too, is always very reliable due to the constituent chlorophyll, which displays a characteristic, intense red fluorescence at an excitation wavelength of 450–490 nm with a 520-nm blocking filter (Fig. 3.99). Materials that do not normally fluoresce can be stained with fluorescent dyes. Figure 3.100 shows how a fluorescent dye is used to show the distribution of adhesive on MDF fibers. In this particular case, the dye was mixed with the adhesive before this was applied to the fibers.

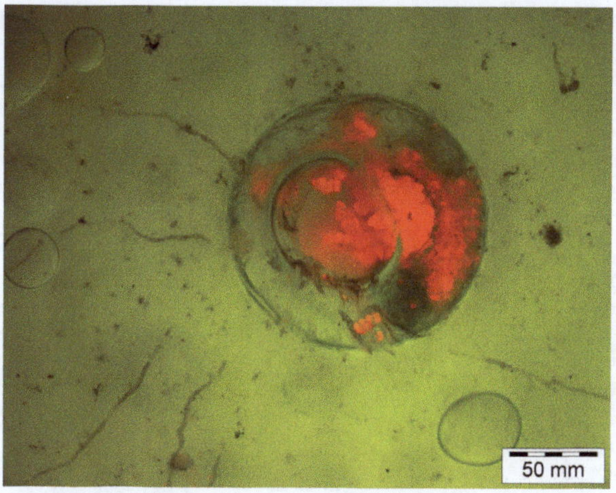

FIGURE 3.99 Upwardly open microfoam-bubbles in coatings provide a sheltered habitat for algae and fungi. At an excitation wavelength of 450–490 nm with a 520-nm blocking filter, the chlorophyll in the algae glows an intense red.

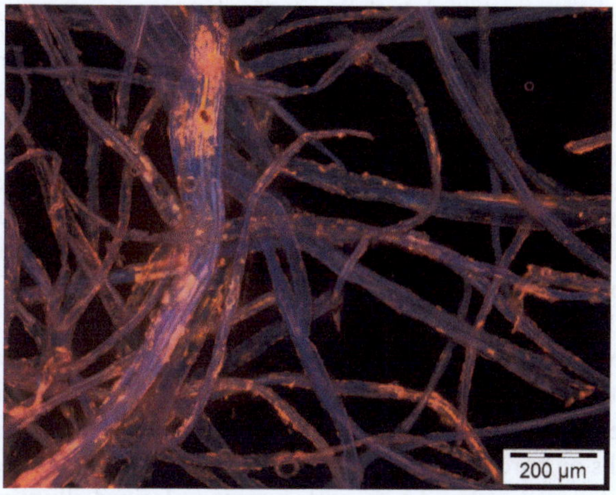

FIGURE 3.100 The adhesive on these MDF fibers was stained with Rhodamine B dye. The adhesive distribution is shown by the reddish fluorescence of the dye under UV excitation (365 nm, with 397-nm blocking filter).

Polarized-Light Microscopy

Some materials change the vibration direction of incident light. If polarized light, that is, light with only a single vibration direction after passing through a polarizer, is used to illuminate a specimen, a second polarizer—generally referred to as an analyzer—which only allows the passage of light with a certain vibration direction can be used to make visible or suppress certain features. If the permitted vibration directions of polarizer and analyzer are at right angles to each other (so-called "crossed polarization"), all materials between polarizer and analyzer that do not change the vibration direction will become invisible.

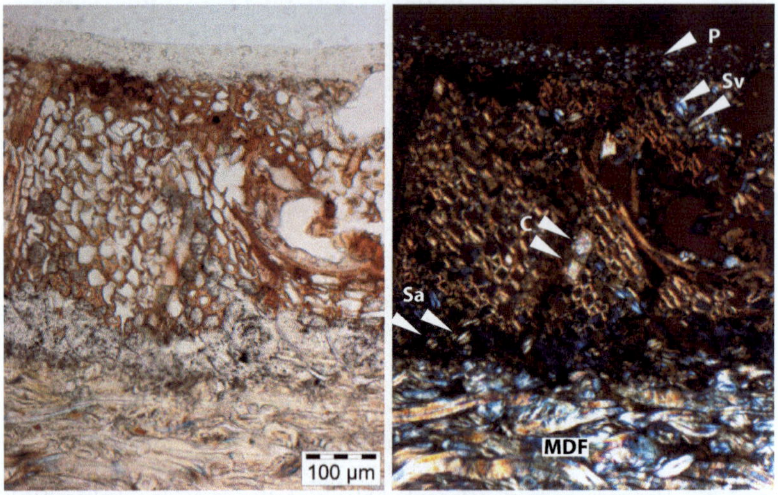

FIGURE 3.101 Same field of view of a coated composite veneer bonded to an MDF panel under normal transmitted light (left) and polarized light (right). The polarized light reveals mineral particles (P) in the coating and crystals (C) in the wood that are typical of the investigated material (obeche). Starch granules are also clearly visible in both the bondline between the veneers (Sv) and the bondline between veneer and MDF (Sa). The MDF fibers are strong polarizers. This is due, among other things, to the compressive strain and microfractures undergone by the fibers during manufacture.

Polarized-light microscopy allows the easy identification of starch granules, mineral particles, crystals in wood and compression lines (Fig. 3.101). It can also provide detailed information on the type and structure of materials.

Cellulose is a particularly strong polarizer on account of its partially crystalline structure. This makes the cell walls of wood easily discernible in crossed-polarized light. Particularly striking are the so-called bordered pits in the cell walls. Intact bordered pits are indicated by a characteristic cross-hatched pattern of birefringence due to the circular arrangement of cellulose. Should the cellulose decay, the polarization effect is lost and the cells or bordered pits appear black in crossed-polarized light (Fig. 3.102). The disappearance of bordered pits under cross-polarized illumination is thus a reliable symptom of brown rot, even in its initial stages.

3.7.4 Film Thickness, Coating System

Failures can result from both overly thick and overly thin coatings. Although some guidelines specify minimum film thicknesses for certain components, such values are of limited use in the evaluation of failure causes. This is because the film thickness needed to protect against particular types of action (e.g., abrasion, weather, or UV radiation) largely depends on the applied coating material. For the practicalities of failure analysis, then, general requirements for film thicknesses should be treated with caution and used only for rough preliminary assessments. The recommendations of the coating manufacturers, on the other hand, tend to be more illuminating. Even here, however, film

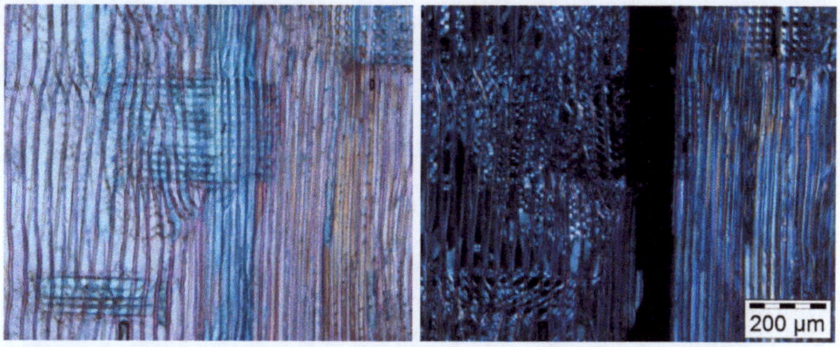

FIGURE 3.102 Pine stained with safranin/astra blue from a piled foundation without (left) and with (right) crossed polarizers. The cells at the center of the image have suffered severe bacterial decay and appear black in crossed-polarized light. Yet, even in the directly adjoining cells, the bordered pits still exhibit the characteristic cross-hatching. The possibility of brown rot infection can thus be ruled out.

thickness specifications pose an inherent danger in failure analysis: given the scarcity of provisions covering other coating properties, the concrete values prescribed for film thickness frequently offer investigators a lifeline that is gladly and unquestioningly accepted. Even if the manufacturer has specified a film thickness between 100 μm and 150 μm and either 70 μm or 200 μm is measured, the case is still far from solved. The deviation from the required value is no more than a piece of evidence which, without further findings, is often of little significance.

Very thick coatings tend to be more prone to defects such as cracking or wrinkling. Coatings are regarded as excessively thin, on the other hand, if their thickness fails to guarantee the required protection. Film thickness has a direct impact on UV transmission, water absorption and water vapor diffusion, making it a particularly important parameter for outdoor finishes.

Various uncomplicated microscopic and non-microscopic procedures are available to determine film thicknesses exceeding around 20 μm on sanded or planed wood. Coatings of less than 20 μm can only be measured by microscope. The smaller the film thickness, the greater is the impact of the natural roughness of the wood surface. At least for films less than 20 μm thick, then, the decision as to how and where to measure needs careful consideration. For coats less than 6 μm thick, the standard measurement method using a reflected-light microscope is often insufficiently accurate to determine film thickness. Such cases require the preparation of transmitted-light specimens along with procedures to minimize the impact of the wood surface on the measurements.

Wedge-cut Drilling

Special drilling devices can be used to cut into films at a very flat, precisely defined angle. The coats can then be measured at a relatively low magnification and the readings converted into the effective thicknesses (Fig. 3.103). This method and two related procedures are described in ASTM D 4138-94.

FIGURE 3.103 Determination of film thickness by wedge-cut method. A built-in magnifying glass and scale are used to measure the distance between the wood and the edge of the coating system at the surface as well as the width of the individual coats. Alternatively, the wedge cuts can also be recorded and measured using a USB microscope. Here too, the measurements can be used to calculate the individual coat thicknesses.

Ultrasonic Measurement

Ultrasonic measurement devices can be used to determine the overall film thickness and, in some cases, the thickness of individual coats. This method is normally suitable for measuring film thicknesses upwards of around 20 μm. However, the accuracies claimed by manufacturers should be viewed with caution, particularly for large-pored woods. Significant deviations in the measured values may also occur with coatings that penetrate deeply into the wood, that is, particularly those with organic solvents. Ultrasonic devices are nonetheless useful, particularly for taking measurements during site visits, provided the results—as with all measurement equipment—are viewed critically (Fig. 3.104).

Number of Coats

As a rule, sequences of different coatings with the same color can be clearly identified using optical or chemical staining techniques. Other useful methods in this regard

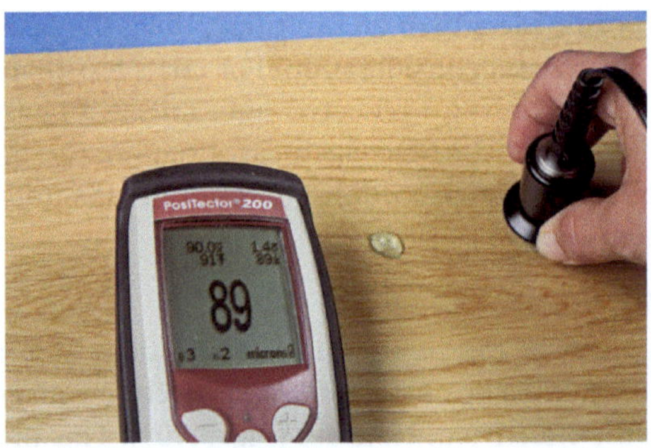

FIGURE 3.104 Ultrasonic measurement of film thickness. A gel is first applied to the substrate and the measurement head pressed onto this.

include fluorescence and general thin-section microscopy as well as iodine fuming on polished surfaces.

In a coating system made up of a single material, on the other hand, identifying the number of individual coats may pose severe problems. With some specimens, the coat boundaries may be indicated by the presence of certain substances that commonly occur at the upper or lower face of the coat. For example, floating waxes viewed under a fluorescence microscope may clearly mark the coat boundaries (Fig. 3.105). Microfoam bubbles are another possible pointer given that they never extend across boundaries. Yet, despite the sometimes easy recognizability of individual coats, it would be wrong to assume that the absence of clear-cut boundaries necessarily implies single-coat application. Investigations conducted on three coating types (acrylic, alkyd and two-component polyurethane) by the Fraunhofer Institute for Wood Research (WKI) in Braunschweig have shown multi-coat application of the same material to be virtually undetectable, irrespective of any intermediate sanding or the drying time allowed between coats. Neither polarized-light microscopy nor fluorescence techniques, using thin sections and reflected-light specimens, revealed the individual coats of the three finish types, applied with or without intermediate sanding, and subject to drying times of up to 24 hours (Fig. 3.106).

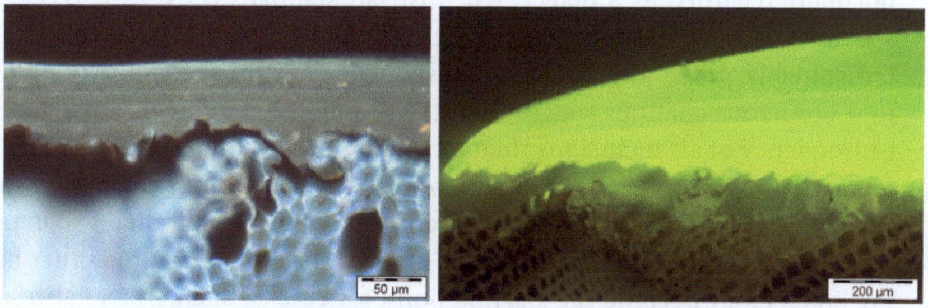

FIGURE 3.105 Fluorescence micrographs of polished surfaces, each showing several coats of same material.

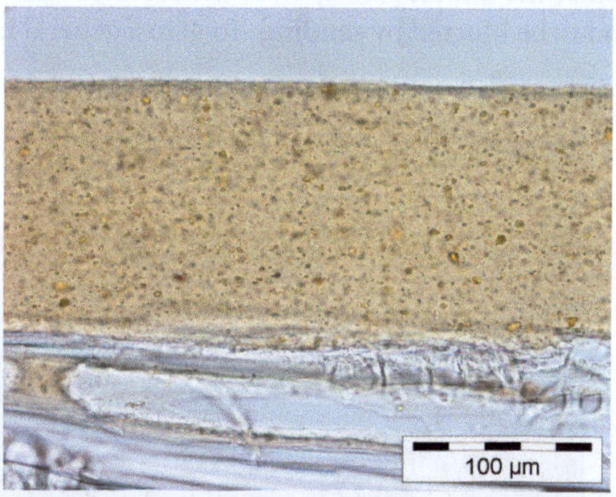

FIGURE 3.106 Acrylic coating viewed under transmitted-light microscope. Although only a single coat can be discerned, the finish actually comprises seven separate coats, applied at increasingly short intervals. Even the bottom coat, which was touch-sanded after 24 hours' drying, is indistinguishable.

Detection of Impregnations

Colorless wood preservative impregnations, whose binder contents range between 3 and 10%, cannot normally be identified by standard microscopic examination. Yet, some are detectable using special illumination or staining methods. The following microscopic methods are suitable for this purpose:

- Fluorescence microscopy
- Staining of thin sections, for example, with Sudan IV
- Staining of dry thin sections by iodine fuming

In some cases, however, wood preservative impregnations resist all efforts to make them visible. The only solution here is a chemical analysis (detection of active ingredient).

Microfoam

The improper proportioning of defoaming agents, incorrect injection parameters or unsuitable drying conditions may result in the entrapment of large quantities of air bubbles (microfoam) in cured liquid coatings. Although some coatings do have the distinction of being practically free of microfoam, the presence of microfoam in coatings is not normally regarded as a defect. An Austrian standard (ÖNORM B 3803) exists which specifies an upper limit for the quantity of bubbles. Yet, there appear to be no studies that indisputably prove any negative consequences of microfoam. My own investigations into numerous coatings suggest that the durability of a finish is affected less by the size or frequency of air inclusions than by their contact with the surface, that is, through the formation of "craters" either during original application or during service (Fig. 3.107). Bubbles that are fully enclosed by the coating material and remain so in use seem to play only a minor role. Where bubbles are open to the surface, on the other hand, the resulting craters may become the starting point for cracks. Such craters also serve as an ideal biotope for surface fungi and algae (see Fig. 3.99).

For the analysis of coatings in cross-section, cleanly cut specimens are needed to recognize whether the bubbles are open to the surface. With softer wood finishes, these fine details tend to be blurred by sanding. To gain a general impression of the frequency

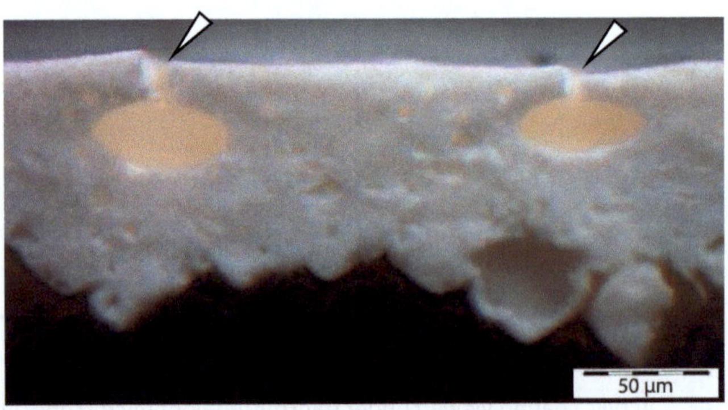

FIGURE 3.107 The cross-section of a coating shows two microfoam bubbles that are open to the surface.

of craters, inspection of the finish surface under glint illumination (see Sec. 3.7.3) or very oblique lighting is particularly recommended (Fig. 3.108).

3.7.5 Analysis of Starch Granules

Starch granules are detectable by microscope in many sapwoods, where their concentration depends on both the tree species and time of felling. Starch is readily stained by Lugol's iodine (see Sec. 3.8) (Fig. 3.109).

Starch, in the form of rye or wheat flour, is also used as an extender in urea-formaldehyde (UF) and melamine-formaldehyde (MF) adhesives. Extenders are additives with their own adhesive action while fillers, such as coconut shell meal, exhibit

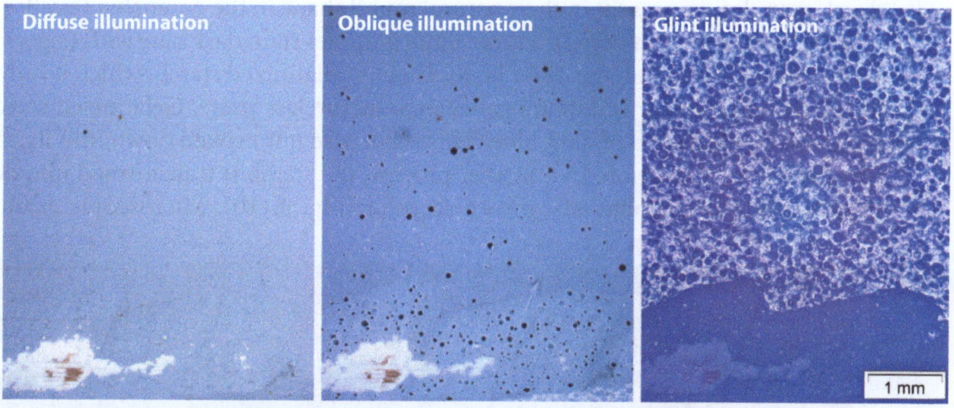

FIGURE 3.108 A razor blade was used to cut into the lower part of the pictured coating at a flat angle. The images demonstrate the widely varying detectability of the microfoam bubbles under the different types of illumination.

FIGURE 3.109 Cross-section through oak with bark. Staining with Lugol's iodine reveals the presence of starch only in the sapwood.

no adhesive properties. Starch additives increase the viscosity of the adhesive, have a positive impact on its structure (shortness instead of stringiness), prevent excessive adhesive penetration, reduce brittleness and shrinkage stresses (with UF adhesives), and help to maintain the required moisture.

As starch granules lose their shape and gelatinize in the presence of heat and moisture, the structure of these granules can occasionally provide evidence of temperature and moisture conditions at an earlier point in time. When heated with water, starch swells at 47–57°C, the granules burst, and starch paste forms at 55–87°C. Starch consists of two components, amylose and amylopectin. When exposed to iodine, amylose stains blue and amylopectin reddish-purple.

With wood products that have undergone little or no steaming, the plant's own starch granules are present in their natural round shape in the parenchyma cells (storage cells for reserve food material). These starch granules stain dark blue with Lugol's iodine. Well-steamed woods, on the other hand, contain gelatinized starch, which is only visible under the microscope as a homogeneous, structureless paste. Gelatinized starch then stains wine-red instead of dark blue. This is because it is broken down into its constituents during gelatinization. In a further process, the starch is transformed into dextrins, identifiable by a characteristic brownish color (Fig. 3.110). Microscopic examination

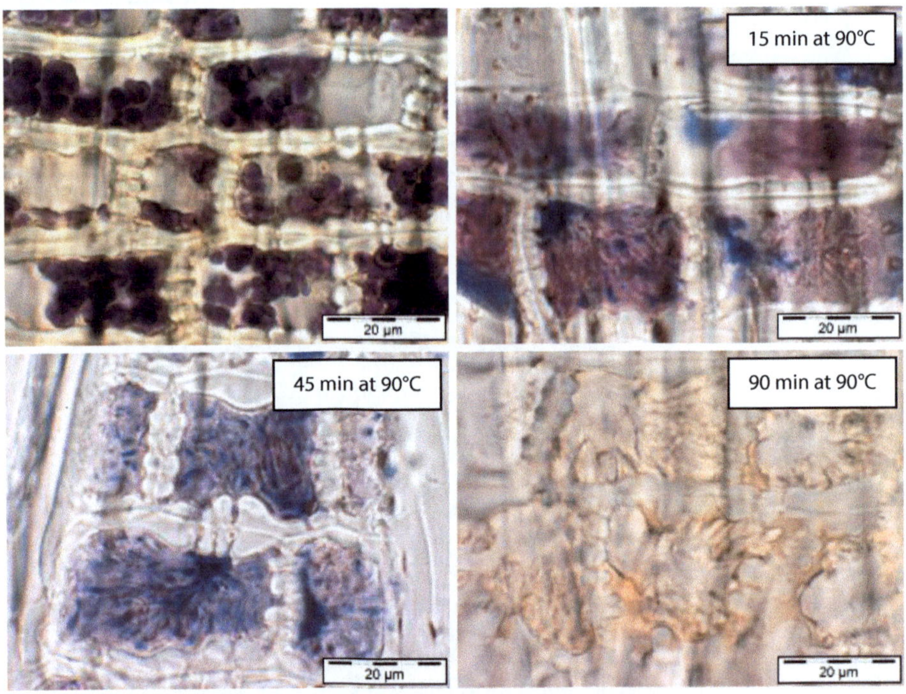

FIGURE 3.110 Starch granules in beech stained with Lugol's iodine. Top left: Original condition. Under continued storage in 90°C hot water, the granules burst and the blue-stained amylose is separated from the red-stained amylopectin. In the course of further boiling, the colors fade and the brown-colored dextrins remain.

can thus provide a clear indication as to whether and to what extent a wood product was steamed.

3.7.6 Microscopic Characteristics in Schematic Form

The sketches in Table 3.2 use the example of a wood-based product faced with a coated hardwood veneer to illustrate typical findings from the microscopic examination of cross-sections.

a.	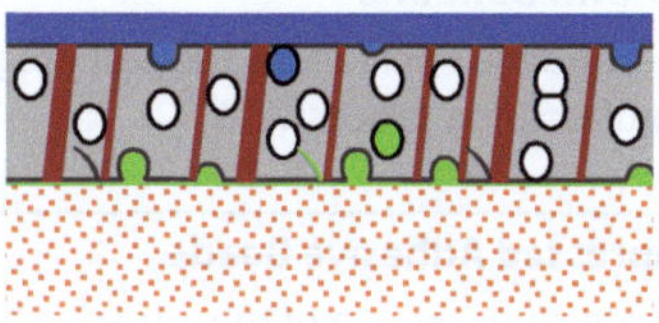	**Best Case** The bondline is thin and uniform, the priming coat has filled the pores. Small lathe checks are filled with adhesive or coating material. There are no regulations specifying the position of lathe checks. With book-matched veneer, the so-called "loose side" (with the checks) alternately faces upwards and downwards.
b.	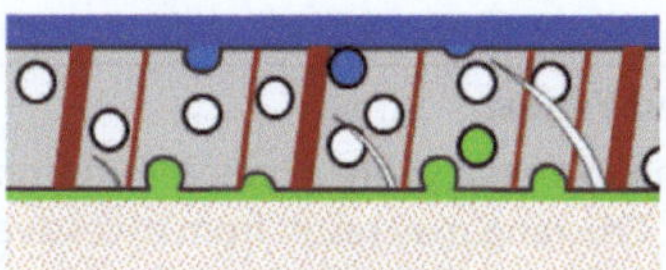	**Lathe Checks** Lathe checks of limited depth are unavoidable and thus have to be tolerated. The general rule of thumb is that checks in veneers used to face furniture items should not have a depth exceeding 50% of the veneer thickness while, in principle, the occurrence of lathe checks in veneer plywood for building purposes is not subject to any restrictions.
c.	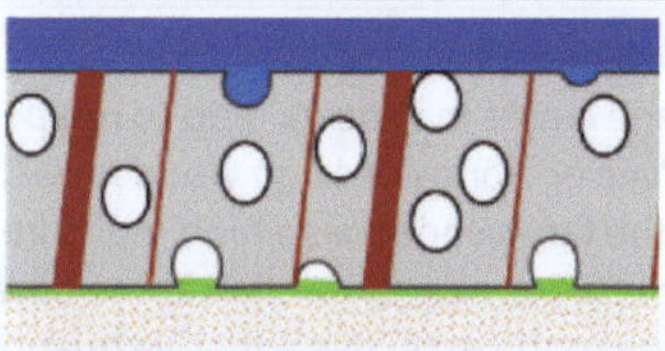	**Pores Not Filled by Adhesive** The adhesive has not penetrated into the large pores of the veneer. The pressure was too low or the adhesive too viscous.
d.	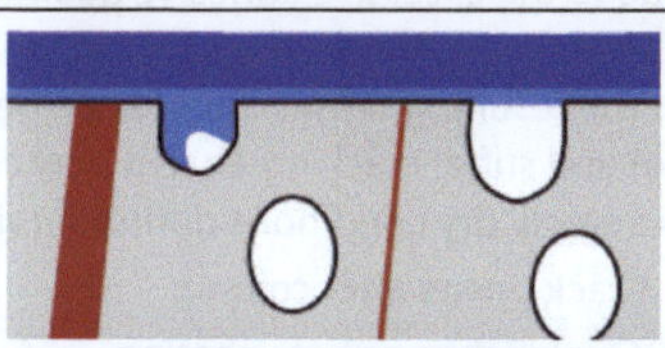	**Pores Not Filled by Priming Coat** The priming coat has not completely filled the pores in the veneer surface. The primer was too viscous or extractives in the pores, for example, may have inhibited wetting.
e.	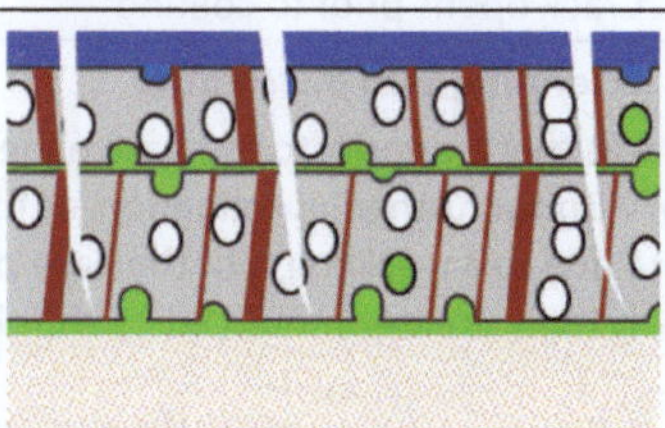	**Parallel Grain Direction of Veneers** The grain direction of the face veneer is parallel to that of the underlying veneer layer. The lathe checks in the thicker bottom veneer have propagated through the bondline and the thinner sliced face veneer.

TABLE 3.2 Microscopic Characteristics of a Wood-Based Product Faced with a Coated Hardwood Veneer

f.	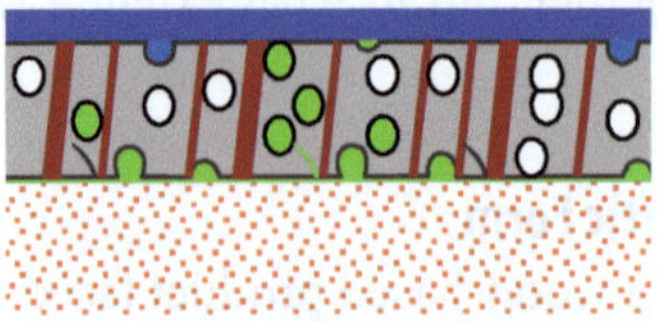	**Adhesive Bleeding** Overly fluid adhesive has penetrated to the top face of the veneer. Adhesive bleeding can cause discoloration even after several years.
g.	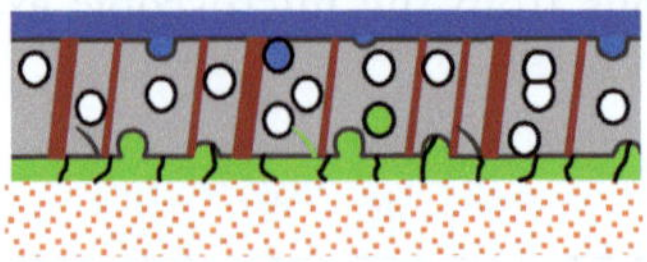	**Thick Bondline** Very thick bondlines can result in delamination, discoloration and cracks in the veneer. When applied in thick coats, urea and melamine resins embrittle and develop shrinkage cracks.
h.	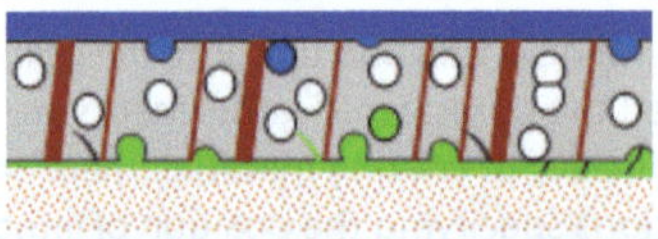	**Nonuniform Bondline** Varying bondline thicknesses suggest irregularities in the manufacturing process.
i.	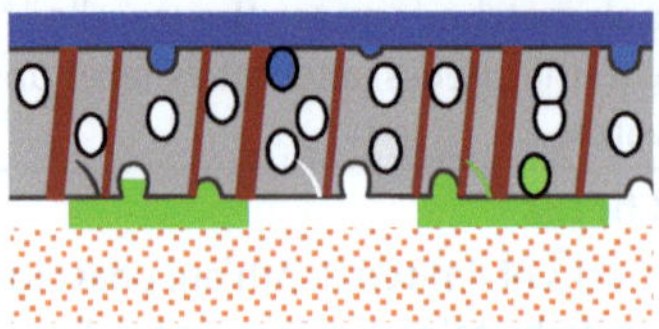	**Uncompressed Adhesive Beads** Inadequate compression of the adhesive beads can lead to delamination, blistering, discoloration and cracks in the veneer.
j.	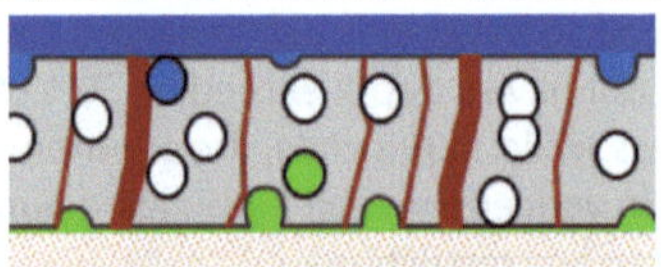	**Crushed Veneer** Deformed wood rays are a typical symptom of crushed veneer. Crushing may result from excessive pressures, particularly in conjunction with high temperature and moisture levels.
k.	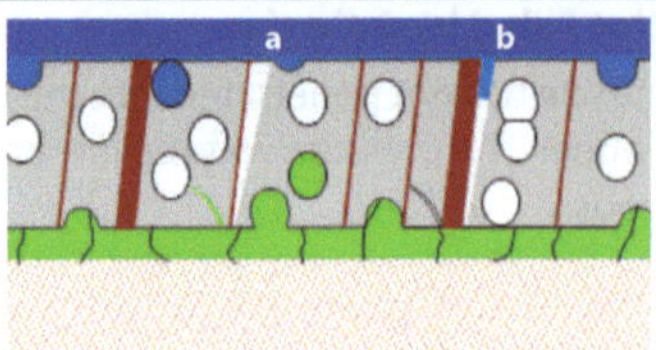	**Shrinkage Cracks** Shrinkage cracks in veneers tend to run perpendicular to the face or alongside large wood rays. They result where the face veneer has a high moisture content and is therefore heavily swollen at the time of bonding (here due to the excessive adhesive quantity). During subsequent drying, the veneers can no longer accommodate the resulting tensile stresses perpendicular to the grain and suffer cracking. Cracks may also result from extremely dry conditions during service. a) Crack opens after coating b) Crack opens prior to coating
l.	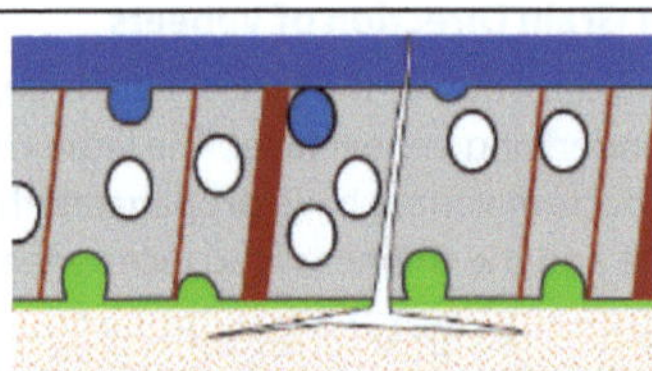	**Cracks Extending into Base Panel** With cracks that extend into the base panel, it remains unclear, without further investigation, as to whether these are due to a weakening of the base panel or extremely high stresses produced by shrinkage of the veneer.

Table 3.2 Microscopic Characteristics of a Wood-Based Product Faced with a Coated Hardwood Veneer (*Continued*)

m.	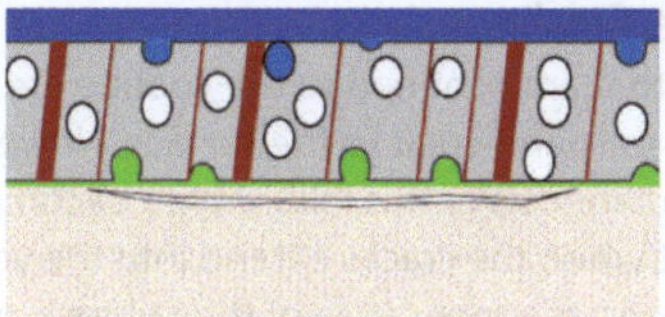	**Detachment Within Base Panel** Where detachment results from a fracture within the base panel, with no cracking observable in the veneer, then the veneer probably played no role in the failure.
n.	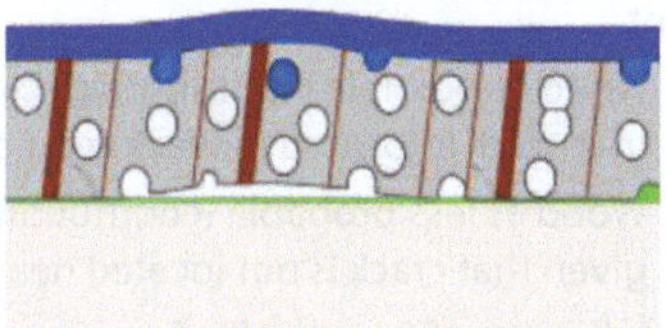	**Blistering of Veneer** The formation of blisters in veneers indicates either that the veneer is more heavily swollen than at the time of pressing or that the base panel has shrunk. Blisters often occur where no contact existed between veneer and adhesive during bonding, resulting in a lack of adhesive penetration during pressing.
o.		**Surface Elevations in Base Panel** Humps in the base panel may be caused by local reductions in surface soundness, large flakes pressed into the surface layer (of particleboard) or localized swelling.
p.	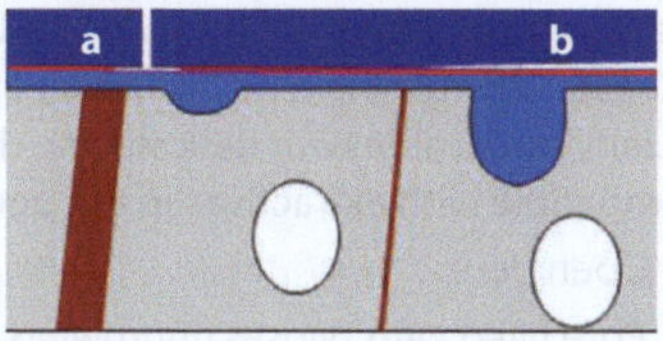	**Intercoat Peeling** Intercoat peeling may be caused by waxes which partly float up to the coating surface during the curing process. These must be sanded away prior to application of the following coat. Failures may take the form of peeling or cracking, depending on the brittleness/strength of the coating. a) Crack with localized peeling (at panel center) b) Peeling (at panel edge)
q.	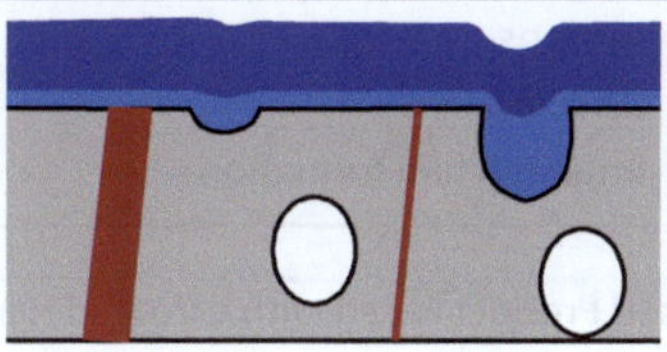	**Telegraphing of Pores** The shrinkage of liquid coatings leads to the formation of small depressions above pores (telegraphing of pores). Particularly with high-gloss coatings, this may impair the visual impact. Possible remedies include longer waiting times between coats, application of several thinner coats or use of a primer with a higher solids content.
r.		**Microfoam** Air inclusions in coatings (microfoam) are, to a limited extent, unavoidable. Large quantities of microfoam lend transparent coatings a slightly hazy appearance. Microfoam bubbles are more or less critical depending on their size and position within the coating.

Table 3.2 Microscopic Characteristics of a Wood-Based Product Faced with a Coated Hardwood Veneer (*Continued*)

s.	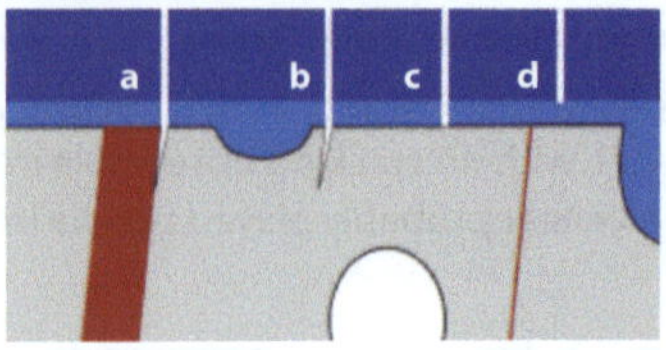	**Coating Cracks** Cracks that are confined to the coating are always exactly perpendicular to the surface. However, microscopic diagnosis of this failure type becomes more difficult when the cracks extend into the veneer. In such cases, the root cause of the failure is not always apparent. a) Wood as possible (contributing) cause given that crack is located in weak zone of wood (ray) b) Wood as less probable (contributing) cause given that crack is not located near rays c) Influence of wood unlikely d) No influence of wood
t.	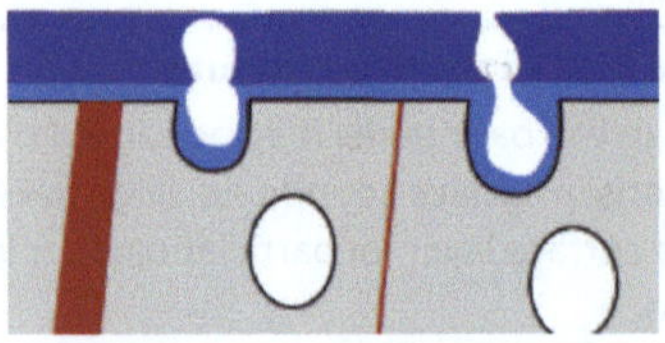	**Air Bubbles Over Pores** Entrapped air bubbles may form above pores in the veneer. These may be enclosed by the coating material or be open to the surface (so-called "pinholes").
u.	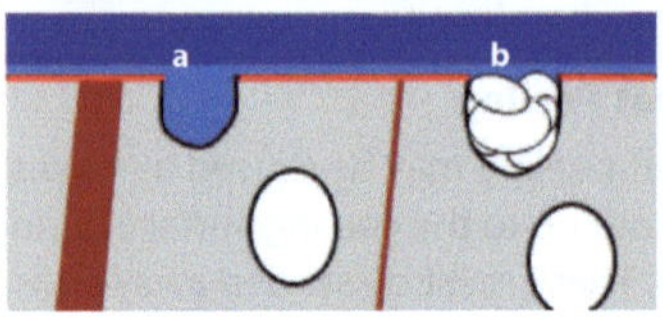	**Pores Without Stain Finish** Where a stain finish does not fully wet the light-colored pores, these will impair the visual impact of the surface (particularly noticeable with dark stains). The stain may be incompatible with extractives in the pores. a) Open pore b) Pore filled with tyloses (ingrowing cell walls)
v.	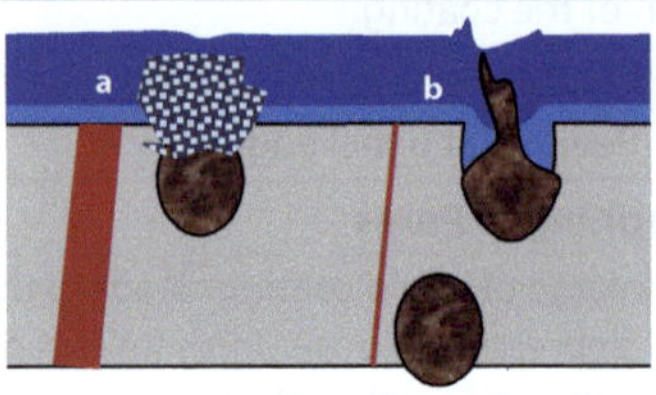	**Extractives in Pores** Extractives present in pores may inhibit film formation or spreading of the coating material and cause haze or color deviations. a) Discoloration b) Impaired film formation

TABLE 3.2 Microscopic Characteristics of a Wood-Based Product Faced with a Coated Hardwood Veneer (*Continued*)

3.8 Chemical Investigation Methods

For all work involving chemicals, it is essential to comply with the relevant safety and disposal regulations!

3.8.1 Staining Methods

Staining methods are based on chemical reactions or physical processes. Some stains are used specifically to detect the presence of a certain substance or substance group, for example, ninhydrin for proteins. Other staining techniques are nonspecific and only reveal that a difference exists between two surfaces. The fact that certain areas of a wood

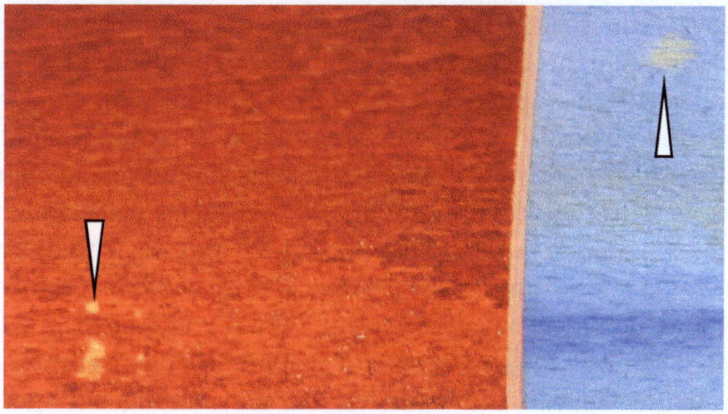

FIGURE 3.111 Discolored patches on stain-finished beech plywood (left) also occurred on untreated material that was "stained" in the laboratory with a blue dye solution. Marks of this kind are a typical sign of contamination, for example, with adhesive.

surface remain unstained when "treated" with diluted ink or an aqueous solution of cotton blue dye, for example, merely indicates that they are unable to absorb the applied material (Fig. 3.111). The actual cause of this, for example, patches of oil, adhesive or another substance, remains initially unclear.

Apart from the use of staining methods for microscopic analysis, which can only focus on small sections, certain techniques can also be applied to solid specimens to establish the distribution of substances over a wider area. This is extremely useful for the clarification of failure causes. For the strict purposes of chemical analysis, staining methods are normally too imprecise and error-prone. Yet, the limitations of any particular staining technique, provided these are borne in mind, are often far outweighed by the findings on the distribution of substances over the specimen.

Some staining methods are suitable for both the direct visual observation of solid specimens and for microscopic examination, while others lend themselves only to one of these applications. The following overview is limited to straightforward, reliable staining techniques that I myself use regularly and successfully. Excluded from the discussion are standard methods using thin sections to investigate wood anatomical features (e.g., safranin—astra blue) and detection reagents for wood preservatives, as described, for instance, in AS/NZS 1605.2. Numerous other staining methods for the detection of particular chemical substances are described in [63] and elsewhere.

The following stains can be brushed or dripped onto the zones under investigation. To highlight the findings, where these are presented in photographic documentation, it may be useful to apply the stain to a small round filter, a cut-to-size paper towel or similar absorbent paper, which is then laid onto the surface (see Fig. 3.119). The boundaries of the stain are then neatly defined by the outline of the paper. It is sometimes useful to place a small upturned glass vessel over the stained area to prevent premature drying out of the liquid. If sections of a specimen have to undergo intensive, long-term staining, the aqueous dye solution can be poured into a small dish that is placed below the area to be treated. The specimen and small dish are then held firmly together and turned upside down—the surface tension of the water preventing any escape of the liquid (see Fig. 3.62). The contents of blisters in coatings can be analyzed by puncturing the blisters with a fine needle and applying suitable dye solutions [68].

Unless the investigator is using very reliable staining methods or techniques with which he or she is highly experienced, the tests should include both negative and positive controls. This requires application of the reagent to specimens which definitely do and do not contain the target substance. In tests for urea with dimethylaminocinnamaldehyde (DACA), for example, the positive control could be performed on particleboard with UF binder. For the negative control, DACA could be dripped onto a fresh piece of solid wood. Only if the particleboard exhibits a color reaction and the solid wood remains unaffected can the stain test on the investigated specimen be considered reliable.

Lugol's Iodine—Starch/PVAc

Lugol's iodine—a solution of elemental iodine and potassium iodide in water—is a convenient, easy-to-use detection reagent that stains both starch and polyvinyl acetate (PVAc). The reagent causes starch to turn blue, red or dark brown to black and PVAc to turn dark red to dark reddish-brown (for microscopy of starch granules, see Sec. 3.7.5). PVAc is the main constituent of most white adhesives. Moreover, some exterior coatings and dispersion adhesives for parquet floors largely consist of PVAc. The characteristic stain colors frequently permit unequivocal identification. The solution is simply brushed or dripped onto solid specimens. The color reaction occurs within a matter of seconds or minutes. (Fig. 3.112) shows the reopened corner joint of a window before and after staining with Lugol's iodine. Figures 3.113 and 3.114 show microscopic applications of Lugol's iodine.

An alcoholic solution of iodine (tincture of iodine) can be used as an alternative to Lugol's iodine. However, the stain produced by tincture of iodine may be too intense in some cases.

Nowadays, povidone-iodine is the preferred iodine formulation for medical purposes. It is not an alternative to Lugol's iodine as it stains only starch and not PVAc, though it is still useful where selective staining of starch in the presence of PVAc is required.

Isocyanate-modified PVAc adhesives are not usually stained by iodine and on very rare occasions, even the staining of "normal" PVAc adhesives is unsuccessful. This may

FIGURE 3.112 Staining with Lugol's iodine of reopened corner joint of a window.

be due to the lack of accessible polyvinyl alcohol, which is probably the target molecule for iodine solutions.

pH Indicators

The results of pH tests may provide immediate clues as to the failure cause, serve to highlight certain material types, for example, adhesives, or help to distinguish the heartwood from the sapwood of meranti and eucalyptus species [86]. The simplest method

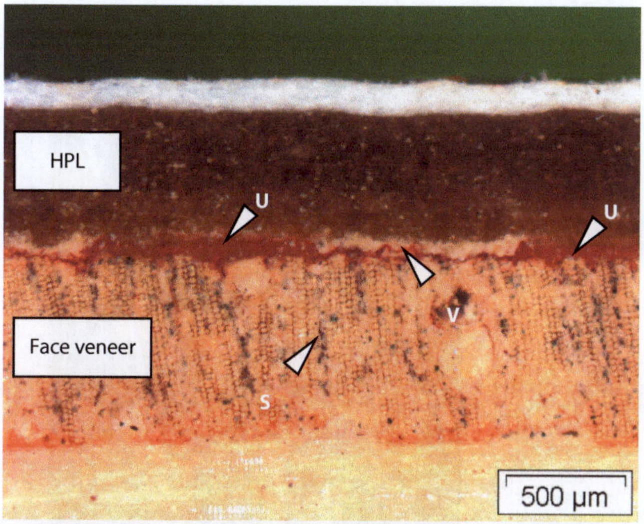

FIGURE 3.113 Polished section of HPL on plywood. The PVAc adhesive was stained dark red by Lugol's iodine. The starch in the face veneer (S) turned black. Clearly observable are the uncompressed adhesive beads (U) and the voids between the beads (V).

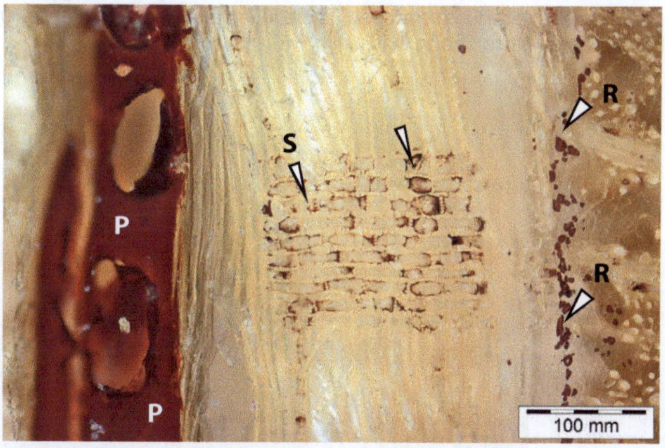

FIGURE 3.114 Cross-section through bonded beech veneers. Staining with Lugol's iodine turned the starch (S) in the wood rays and the rye flour (R)—used as an extender in the UF adhesive bondline between veneers—a brownish color. The PVAc adhesive (P) bonding the veneer to the panel stains reddish-brown.

involves the use of standard commercial pH indicator papers. Water is dripped onto the surface under investigation and left to stand for a short time. The pH indicator paper is placed onto the surface just before the moisture disappears (Fig. 3.115).

Universal indicator solution is a proven means of highlighting adhesive joints and of documenting the distribution of pH values [29]. The solution contains several pH indicators that allow adhesive to be easily distinguished from wood, provided the adhesive and wood have different pH values (Fig. 3.116). Variations in the pH value across a

FIGURE 3.115 Determination of pH value with indicator paper. The pH value of 5 recorded for the surface of the varnished oak veneer indicates only weak acidity. A strongly acidic pH value of 1 was measured where the surface was scratched. The constituents of certain fire-protective coatings have a tendency to break down into strong acids.

FIGURE 3.116 Staining of melamine-urea-formaldehyde (MUF) adhesive with universal indicator solution. The red stain of the acidic adhesive contrasts with the green stain of the wood.

surface due to the action of acids (Fig. 3.117) or fungal attack (Fig. 3.118) can also be documented in this way.

The indicator comprises 0.25 g methyl orange, 0.25 g bromocresol green, and 0.01 g indigo carmine, which are dissolved in 50 mL ethanol and topped up with 25 mL of deionized water.

In addition to the pH value, the so-called "buffering capacity" of materials, that is, the speed with which the pH value changes upon the addition of acids/alkalis, can be measured with pH electrodes on small pieces of the material after storage in water for a predefined period.

Cotton Blue—Various Applications

Cotton blue is an easy-to-use, water-soluble blue dye that is suitable for a wide range of applications. Tests performed to date have shown that the dye front advances at roughly the same rate as the water front. Cotton blue stain thus offers a convenient means of visualizing the infiltration of water into wood (see Fig. 3.61). Where penetration behavior is not the focus of the investigation, diluted fountain pen ink can be used as an alternative, for example, to make visible the grain direction (see Fig. 5.10).

Potassium Thiocyanate—Iron

Potassium thiocyanate/potassium rhodanide can be used to identify iron as the cause of discolorations in wood or coatings. When dissolved in water together with a small

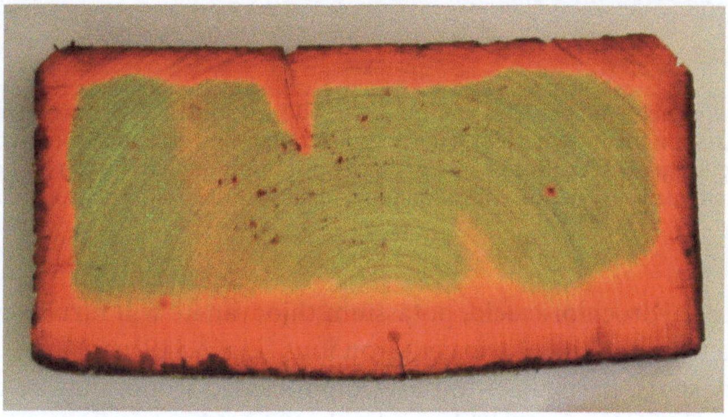

FIGURE 3.117 Cross-section of a beam from a hot-dip galvanizing plant after treatment with universal indicator solution. The outer zones of the timber have been attacked by hydrochloric acid vapors.

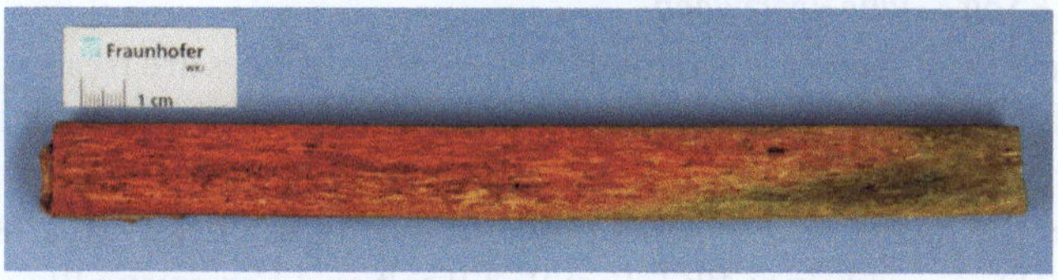

FIGURE 3.118 The shift in the pH value reveals the extent of fungal attack in a cross-section of particleboard. The red zones are infected by fungi.

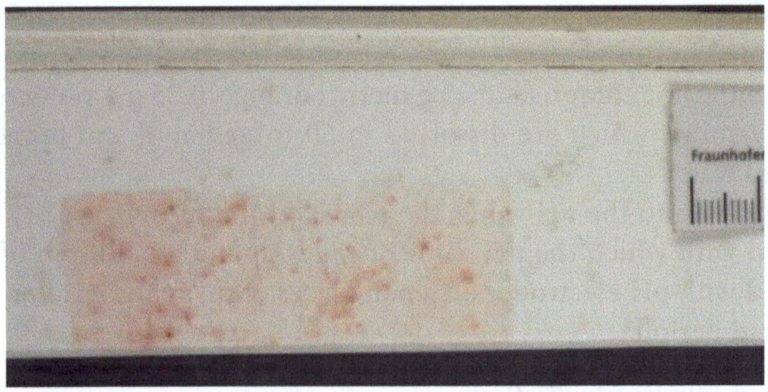

FIGURE 3.119 Gray discoloration on a painted window surface. A potassium thiocyanate solution is drizzled onto a piece of absorbent paper, which is then placed on the window. The red stain shows iron to be the cause of the discoloration.

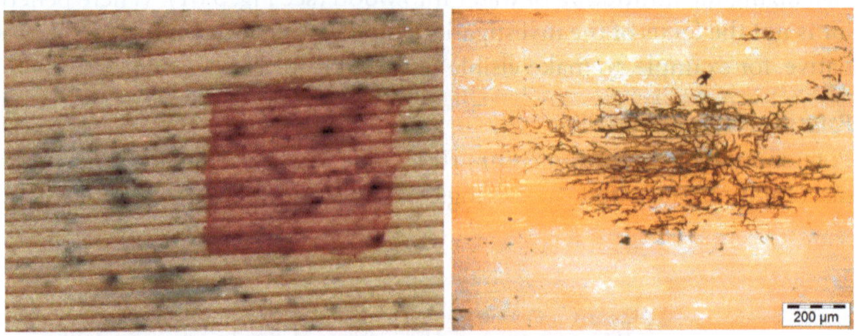

FIGURE 3.120 Left: Iron particles on Douglas fir. The red square shows the positive detection of iron using potassium thiocyanate. Right: Black spot on larch viewed under microscope, revealing presence of staining fungi. As a rule, these are not clearly distinguishable from iron particles with the naked eye. However, simple examination with a magnifying glass usually allows correct diagnosis.

amount of hydrochloric acid, potassium thiocyanate will turn deep red in the presence of even a tiny quantity of trivalent iron ions (Fig. 3.119).

Black spots, particularly on raw, unprocessed wood, may be caused by either iron particles or staining fungi. This can be clarified either by an iron-detection test or examination under a magnifying glass or microscope to see if fungal hyphae are present. Filamentous structures are a reliable sign of fungal attack (Fig. 3.120).

Fast Blue B Salt—Pine Heartwood

With some pine species, particularly where the surfaces are freshly worked, it is virtually impossible to distinguish the heartwood from the sapwood with the naked eye. Brush or spray application of a fresh solution of approximately 0.2% Fast Blue B salt in water will stain the heartwood of Scots pine, for example, a deep red after only a few minutes (Fig. 3.121). According to [86], Fast Red B salt and Variamine Blue RT salt (described in AS/NZS 1605.1) are also suitable for detecting the heartwood in various pine species. As yet, however, no satisfactory detection method has been found for spruce heartwood.

FIGURE 3.121 Heartwood of Scots pine stained with Fast Blue B salt.

Sudan IV—Fats, Waxes, Oils

Sudan IV can be used to stain hydrophobic (water-repellent) substances, such as fats, waxes and oils, as well as many hot-melt adhesives. (Other Sudan dyes and Nile Blue Sulfate, for example, are also suitable.) Thin sections are very easy to stain: Sudan IV is dissolved in ethanol (e.g., 0.1 in 80% ethanol) and drizzled directly onto the surface. Washing of the specimen is not necessary prior to examination.

The staining of solid specimens with Sudan IV, on the other hand, is a much more elaborate process. Here, successful testing depends on the use of special procedures to ensure selective diffusion of the dye into the target substance.

One option is to soak the entire specimen in a solvent compatible with the dye, for example, ethanol. The subsequent addition of Sudan IV to the impregnating solution will then result in a selective accumulation of the dye at those places where hydrophobic substances are present. The solvent must then be allowed to evaporate. However, as prolonged contact with ethanol may cause some materials to disintegrate, this method can only be recommended in rare cases.

An alternative option is to apply the dye through selective diffusion from a solid. The simplest method involves sprinkling pure dye powder or a mixture of dye powder and ground chalk onto the specimen surface. The hydrophobic zones in the surface will stain red after several hours at room temperature or a few minutes at a higher temperature (Fig. 3.122).

A much finer contrast can be achieved through the use of stained putty. Here, a permanently elastic putty is stained by kneading powder into a white putty product (e.g., Terostat VII—Henkel). The stained putty is then pressed onto the finely sanded surface of the specimen. Within a few hours at 60°C or one or more days at room temperature, the stain will diffuse into all hydrophobic substances in the surface and highlight these in red. To gain a better contrast, gentle sanding of the surface after staining may be useful (Figs. 3.123 and 3.124).

DACA—Urea and Polyurethanes

Dimethylaminocinnamaldehyde (DACA) stains urea and UF resins [78] (Figs. 3.125 and 3.126) as well as polyurethanes (see Fig. 3.17). Although a solution of 0.3% DACA

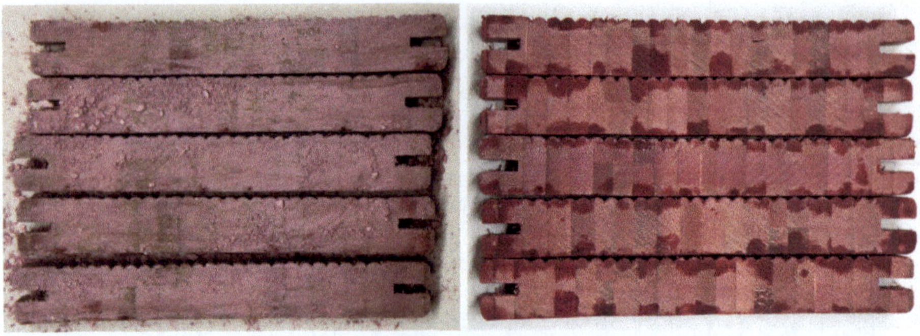

FIGURE 3.122 In order to document the penetration depth of an oil-based wood preservative, Sudan IV powder mixed with chalk (1:20) was applied to the cross-sections of the decking boards. After one hour's storage of the specimens at 60°C, the powder was removed to reveal the penetration depth.

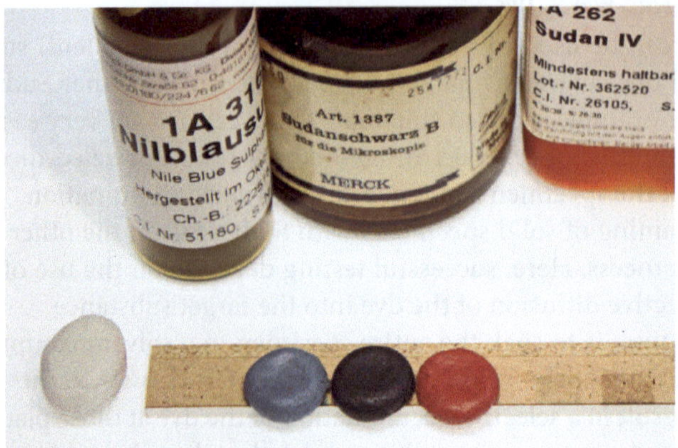

FIGURE 3.123 Staining with Nile Blue Sulfate, Sudan Black and Sudan IV using stained putty pressed onto cross-section.

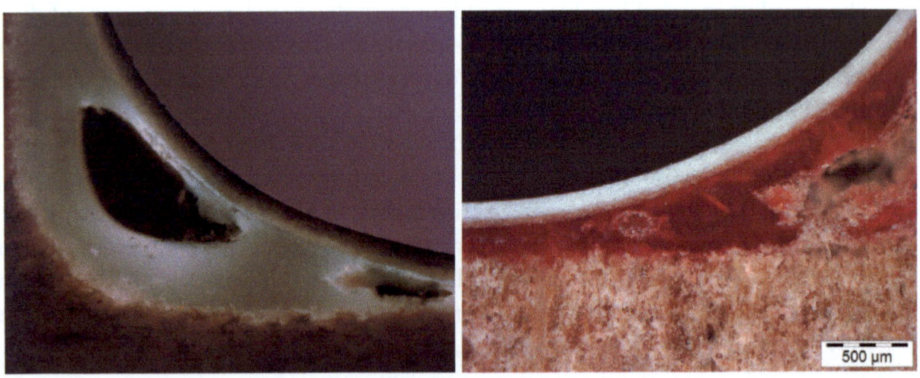

FIGURE 3.124 Hot-melt adhesive stained with Nile Blue Sulfate (left) and Sudan IV (right). With Nile Blue Sulfate, the adhesive is mainly highlighted by the resulting fluorescence. Sudan IV stains the hot-melt red.

FIGURE 3.125 Staining of three-layer parquet with DACA. The urea-formaldehyde (UF) adhesive exhibits a reddish stain. Checks in the wood are also filled with adhesive. This shows that the checks were already present at the time of bonding and that the pressure cannot have been too low.

FIGURE 3.126 Staining with DACA causes the UF adhesive in laminated veneer lumber to turn a reddish color. The dye also stained the beech veneers green.

in ethanol/15% hydrochloric acid (5:2 V/V) can be used for around two months, the color reaction becomes weaker the longer the storage period.

The speed of staining is temperature dependent. At room temperature, a change in color can be expected after 2–15 minutes. The stain is unsuitable for use with transmitted-light microscopy and for the identification of very fine structures as it always extends into the adjacent areas.

Some hardwoods, including beech and cherry, are stained green by DACA (Fig. 3.126). While the exact cause of the reaction is unknown, certain tannins are thought to be responsible [78].

Ninhydrin—Proteins

The reagent ninhydrin is used to detect proteins. 0.1 g ninhydrin is dissolved in 10 mL of 96% ethanol and topped up to 100 mL with deionized water. Possible applications include the detection of fingerprints [73], urine, contamination with plant seeds, protein adhesives and paints, as well as the visualization of protein distribution in wood. Figure 3.127 shows how staining with ninhydrin reveals the presence of linseeds in a particleboard surface. Figure 3.128 illustrates use of the stain in estimating the protein content of bamboo strips.

Iodine Fuming—Various Applications

Iodine fuming—as a well-established forensic technique [73]—occasionally proves useful in failure analysis. Although iodine fumes only show a characteristic reaction with a

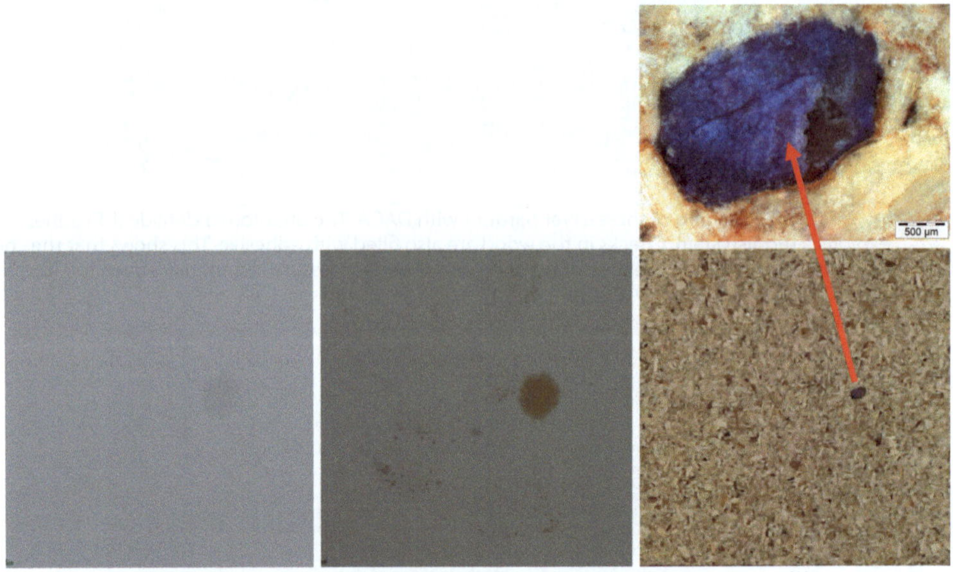

(1) (2) (3)

FIGURE 3.127 Discolorations on backing paper (1) were stained by iodine fuming (2). After sanding away the paper, the underlying linseed was stained blue with ninhydrin (3).

FIGURE 3.128 Two adjacent sections of a panel with a bamboo strip core. The top section was stained with Lugol's iodine. The dark-colored zones show the occurrence of starch in the bamboo strips. The bottom section was stained with ninhydrin. Here, the purple coloration of individual strips indicates the presence of proteins. No correlation was thus observed between the occurrence of starch and proteins.

limited number of substances, the typical color differences between individual layers or materials provide a helpful visual contrast between otherwise near-indistinguishable materials. Fuming, as such, also has a number of fundamental advantages over liquid staining:

- The risk of substances being leached by the solvent, for example, water, is eliminated.
- No swelling of the wood will occur.
- The progress of staining can be visually assessed and the fuming test aborted or extended as required.

For the fuming test, a small amount of iodine is placed in the bottom of a transparent vessel. The specimen is placed such that exposure of the area under investigation to iodine fumes is as uniform as possible, with due allowance for the fact that the fumes are heavier than air (Fig. 3.129). The specimen is removed as soon as the stain has achieved an adequate intensity. Appropriate exposure times generally range between 15 minutes and two hours, though the reaction can be significantly accelerated by heat. Excess iodine should then be allowed to evaporate for a short time prior to microscopic examination. Due to the toxicity and corrosive action of iodine, the test must be performed under a fume hood. The staining can be assessed by visual or microscopic examination.

Iodine tends to accumulate in low-polarity substances (e.g., the fatty components of fingerprints). It undergoes a chemical reaction with unsaturated fatty acids, for example, in linseed oil (see Fig. 3.127), and has been shown to produce an intense stain in the presence of some polyurethane adhesives (Fig. 3.130). Moreover, it evidently displays a very specific reaction with epoxy resins, which turn a strong red when fumed (Fig. 3.131). Pitch can also be reliably detected by iodine fuming (Fig. 3.132). The fact that iodine-fumed surfaces lose their fluorescence can sometimes be helpful in diagnosing failures (Fig. 3.133).

FIGURE 3.129 Iodine fuming of specimen. The specimen is placed with elemental iodine in a transparent sealable vessel. It is removed as soon as the stain has achieved an adequate intensity.

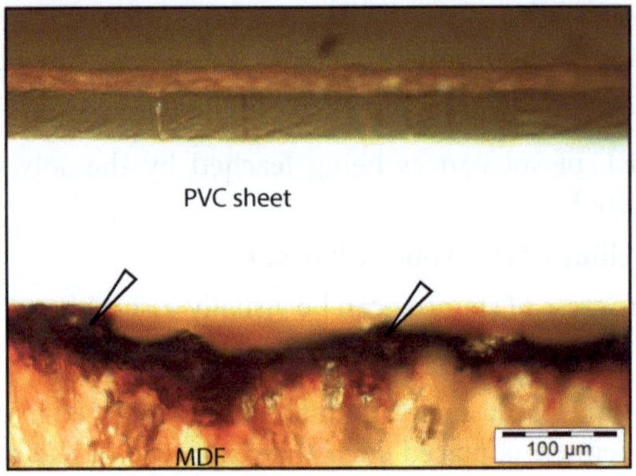

Figure 3.130 Polyurethane adhesive between medium-density fiberboard (MDF) panel and PVC sheet, stained dark brown by iodine fuming.

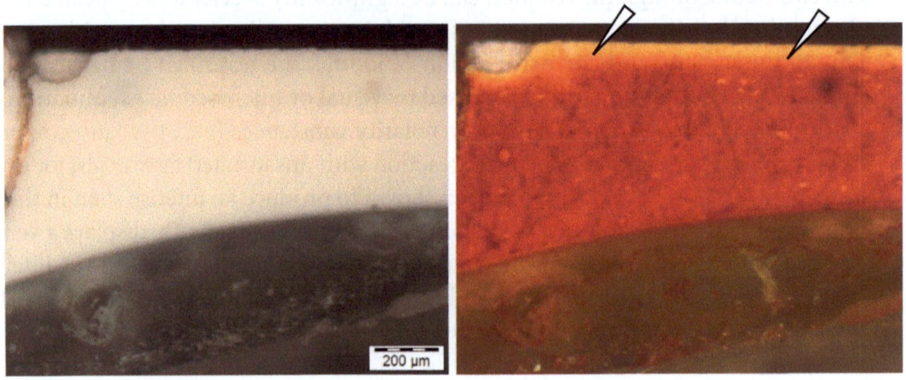

Figure 3.131 Cross-section of weathered epoxy resin surface before (left) and after (right) iodine fuming. The less intense stain in the upper zone indicates that the epoxy resin here has degraded.

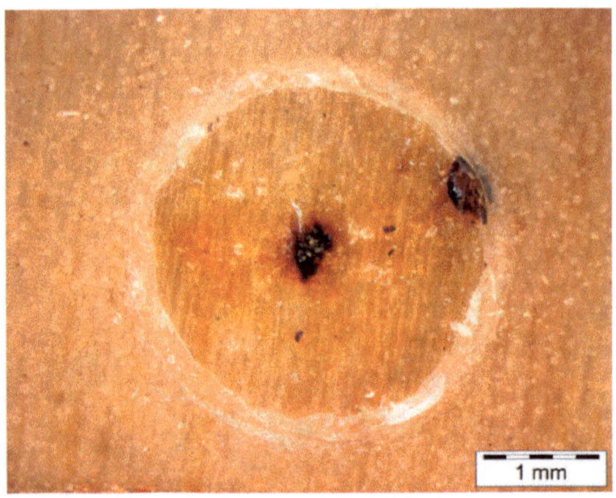

Figure 3.132 The brown stain produced at the base of a blister by iodine fuming shows that pitch from the wood has migrated into the priming coat in this area.

FIGURE 3.133 Cross-section of varnished veneer (wood species sapele) on particleboard. The analysis set out to establish whether adhesive bleeding was the cause of discolored patches on the top face. The two images show the identical cross-section before (top) and after (bottom) iodine fuming under UV fluorescence excitation. Prior to fuming, both the varnish and adhesive exhibit blue fluorescence. With this specimen, the iodine is primarily absorbed during fuming by the transparent varnish. Given that iodine suppresses the fluorescence of the varnish, the test clearly demonstrates that adhesive has bled into some of the near-surface vessels.

3.8.2 Overcure and Undercure of Melamine Resins

Melamine resin-impregnated paper is one of the most widespread particleboard and fiberboard finishes. The impregnated papers are either fully cured during manufacture and bonded to the board surface with an additional adhesive in a separate operation (continuous-pressure laminate/CPL or high-pressure laminate/HPL) or directly fused onto the surface of the wood-based product, with the melamine resin doubling up as adhesive (melamine-faced board/MFB or direct-pressure laminate/DPL).

Melamine resins cure under pressure and at high temperatures. During this process, chains of resin molecules cross-link with each other to form a network that offers the desirable properties of hardness, chemical resistance, etc. As the brittleness of melamine resin and, with it, the tendency to certain types of cracking increases in line with the degree of cross-linking, too much cross-linking is ultimately a disadvantage. It is therefore necessary to control the reaction so as to ensure adequate curing while preventing an excessive amount of additional cross-linking (overcure).

In practice, the so-called Kiton or acid test, as described, for example, in AS/NZS 4266.21, is frequently used to determine the degree of cure. This involves application to the surface of sulfuric acid with a red dye. After two hours, the acid is removed and the surface assessed in terms of stain intensity and change in gloss level. It is assumed that the resulting color will be proportional to the degree of resin cure. A surface that shows hardly any signs of acid attack and has absorbed virtually none of the dye is accordingly classed as overcured.

Yet, the observed change in surface color, in particular, may invite serious misinterpretation where paper fibers are present near the surface. This is due to the impossibility of distinguishing between the dye absorption of the resin and the staining of the paper fibers. Assessment of the stain is in any case hardly feasible with dark melamine finishes. Hence, while the method may be suitable for the purposes of production monitoring—where the adopted resins, papers and procedures are familiar—it is seldom useful in the analysis of failures on surface finishes of unknown origin.

As an alternative to visual assessment of the acid test, the amount of melamine hydrolyzed by the acid may be quantified by its UV absorption. Due to its interference with UV absorption, sulfuric acid has to be replaced by hydrochloric acid for this procedure. The hydrochloric acid is applied to the surface for a predetermined period and then transferred to a UV spectrometer, where the UV absorbance at 237 nm is recorded.

As different resin formulations deliver different results irrespective of their curing status, the values obtained by this method are useful only for comparative purposes. The values may be compared to older or younger material from the same production line. The other possible comparison is with the same material after heat treatment. In the case of overcure, the melamine content will hardly be affected by heat treatment. Surfaces with the required degree of curing, on the other hand, will show a sharp drop in melamine in the acid after heat treatment.

3.8.3 Instrumental Methods

In crime movies, forensic experts are typically shown putting substances into a piece of equipment and almost immediately reading off the results from the computer screen. Real-life analyses, however, tend to be more complex: even selection of the most suitable technique from the multiplicity of options presently available can be far from easy. Also, with many methods where the measurements are quick to take and unspectacular, the hardest task is to interpret the results. Virtually all common adhesives and coatings are mixtures of several substances. It is quite possible that the critical substances are only present in minute quantities and very difficult to detect. Indeed, from the perspective of chemical analysis, wood is an exceptionally "dirty" material with innumerable constituents in fluctuating proportions. However, as long as the investigator appreciates these difficulties and has at least a basic familiarity with the various analytical techniques, instrumental methods still lend themselves to productive use in routine failure analysis. Nowadays, numerous laboratories offer chemical analysis services at often astonishingly low prices. Normal investigators, like any other professionals, can submit specimens for analysis. It is nonetheless essential to specify that the laboratory should not only deliver the basic results, but also interpret these in the context of the issues under investigation. It is also fundamentally important to accept that not every analysis will deliver a clear-cut answer. Table 3.3 summarizes a number of established instrumental methods of chemical analysis. Two of these, infrared (IR) and energy-dispersive X-ray (EDX) spectroscopies, which are used fairly frequently for special failure analysis, will be discussed in greater detail.

Infrared Spectroscopy

Infrared (IR) spectroscopy is one of the simplest and most effective instrumental methods of chemical analysis [28, 79]. When exposed to radiation in the IR region of the

Method	Analysis	Examples
IR-(ATR) Infrared spectroscopy, often used in conjunction with the attenuated total reflection technique	Chemical "fingerprinting" of substances Comparison with spectra from database Sensitivity not very high and depends on type of chemicals Requires specialist laboratory with large database for relevant substances	What type of adhesive or coating is present? Was curing agent correctly proportioned? Are there significant differences between batches? Are any surfaces contaminated with substances such as oil?
EDX Energy dispersive X-ray spectroscopy	Elemental composition of salts, paints, contamination No details of chemical bonding	What caused efflorescence? Does coating contain particles of a certain composition?
DSC Differential scanning calorimetry	Detection of chemical or physical reactions at certain temperatures Reactivity, melting point, glass transition point	Is a resin still reactive? Does substance A interfere with curing of substance B?
GC/MS Gas chromatography/ mass spectrometry	Detection of light organic substances in liquid or gas phase	Does coating or adhesive contain plasticizers? Is yellow discoloration caused by nicotine? Are there any differences in chemical composition of paraffins of two particleboards?
TOF/SIMS Time-of-flight secondary ion mass spectroscopy	Very sensitive detection of atoms and molecules at surface Very careful handling required to avoid contamination	Is any contamination present in boundary layer of failed bond? What is cause of discoloration? What additives are present in polymers?

Table 3.3 Established Instrumental Methods of Chemical Analysis for Use in Failure Analysis of Wood and Wood-Based Products

electromagnetic spectrum, the tested substances absorb characteristic wavelengths. Specific absorption bands reveal the existence of certain parts of molecules while other absorptions can be interpreted as a chemical fingerprint. By comparing the resulting absorption spectra to those of reference spectra from databases or reference materials tested in parallel, it is possible to establish the type of adhesive or coating within a matter of minutes. While certain chemical compounds are very easy to detect by means of IR spectroscopy, others remain invisible even in relatively high concentrations.

Attenuated total reflectance infrared (ATR-IR) spectroscopy is the most frequently applied IR spectroscopy method. Here, specimens are pressed onto a diamond without

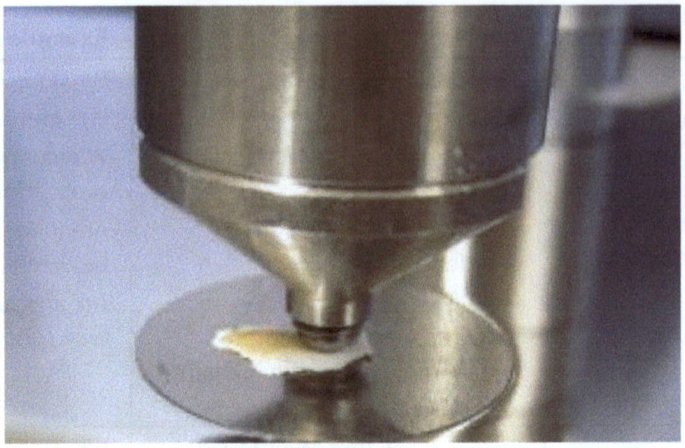

Figure 3.134 A coating fragment is pressed onto a diamond by the pressure tip of the ATR unit.

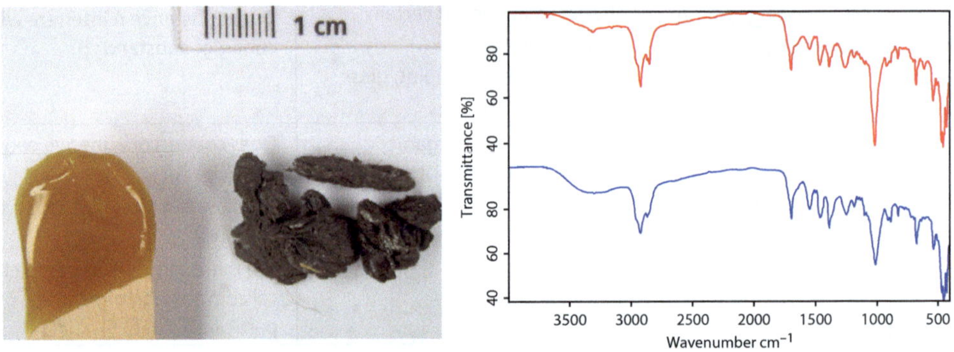

Figure 3.135 The almost identical IR spectra of a honey-colored wax and black residue from flooring show that the two materials have the same chemical composition.

further preparation (Fig. 3.134). Part of the IR radiation reflected in the diamond penetrates into the surface of the specimen, where it is reflected or absorbed depending on the constituent molecules.

Typical questions that can be resolved by IR spectroscopy include:

- What type of adhesive or coating is present?
- Was the curing agent correctly proportioned?
- Are there significant differences between batches (Fig. 3.135)?
- Are any surfaces contaminated with substances such as oil?

EDX

Energy-dispersive X-ray (EDX) spectroscopy relies on the excitation of electrons in the specimen under a SEM. This excitation causes electrons to emit X-rays with an energy content specific to the particular element. These X-ray emissions thus provide clues as

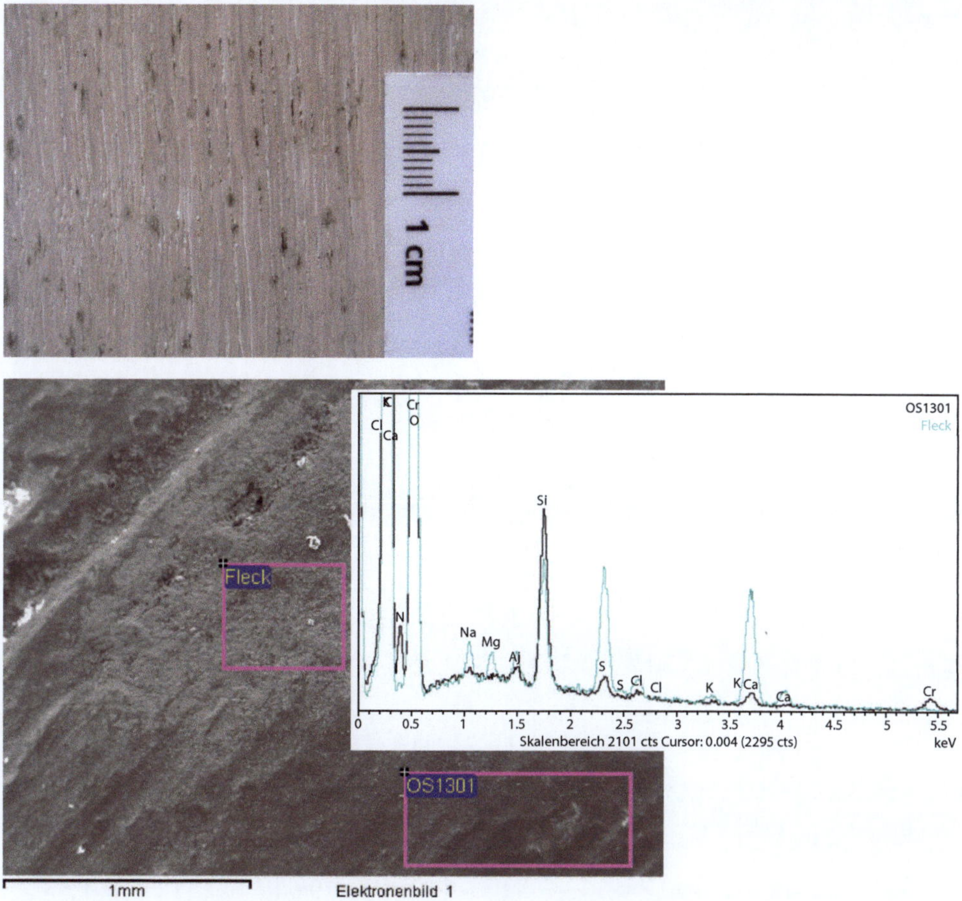

FIGURE 3.136 EDX spectroscopy analysis of dirt in pores of oak parquet. The "Fleck" spectrum is from a dark spot on the parquet, the OS 1301 spectrum from a clean part of the surface. The EDX shows the spot to exhibit much higher levels of sulfur, calcium and oxygen and significantly higher levels of sodium and magnesium, compared to the clean area. This chemical composition suggests that gypsum (calcium sulfate) is the main cause of the discoloration. The most likely source of gypsum contamination in buildings is the residue from construction works (most probably fragments of plaster or gypsum board).

to the elemental composition of the specimen. However, the fact that EDX spectroscopy registers only the type and quantity of the atoms present, without any indication of their bonding, makes it unsuitable for the analysis of organic compounds. It is therefore used to investigate inorganic compounds (e.g., metals and salts) or to detect foreign atoms in organic compounds (Fig. 3.136).

CHAPTER 4

Interpretation of Features and Distributions

All features exhibited by a specimen are a source of information about the manufacturing conditions, materials used or service conditions. In some cases, the absence of particular features can be equally revealing. What does experience suggest to be the most likely features under certain service conditions? What distribution of features would rule out a specific failure cause? Is there any indication of the earliest or latest point in time when the failure may have occurred? What investigation method would substantiate or disprove the initial suspicion based on the failure characteristics or distribution? This section of the book sets out to sharpen the awareness for possible interpretations and demonstrate the value of observations such as those listed above.

4.1 Chronology of Findings

Was there any cracking or bonding failure prior to use? Findings that allow the cause of a defect or feature to be narrowed down to a time before or after a particular production step or period of use are extremely valuable for failure analysis and should always be followed up very closely. One particularly useful phenomenon, described in Sec. 4.1.2 as the "honey-on-toast" effect, has so far received little attention in the relevant literature, even though it clearly links failures to a certain production stage with astonishing success.

4.1.1 Discoloration and Layer Sequences

A layer of dirt or dust on parts of an otherwise fresh fracture offers clear evidence that a crack existed before the failure (Fig. 4.1). Similarly, the presence of coating material in a crack proves that the crack predated the finish application (Fig. 4.2). Even a small paint splash running across a crack can provide clues as to when the cracking occurred (Fig. 4.3). It is generally reasonable to assume that more recent layers will overlie older layers (Fig. 4.4). While the sequence of layers in a coating, etc., is normally established

FIGURE 4.1 The dark patches on parts of the fracture surface of a square-sawn timber indicate that cracks running across the grain existed prior to the fracture.

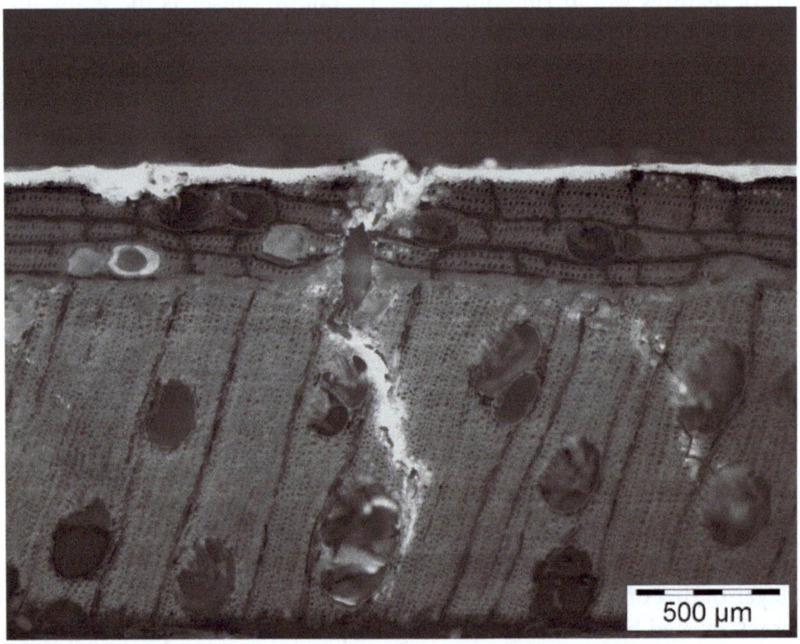

FIGURE 4.2 Cross-section through plywood used as siding on facade. The fact that coating material has penetrated into the lower veneer layer proves that the crack was already open when the last finish was applied. Veneers bonded together with their grain direction parallel, as illustrated here, typically suffer heavy cracking as lathe checks can easily propagate from one veneer into the other.

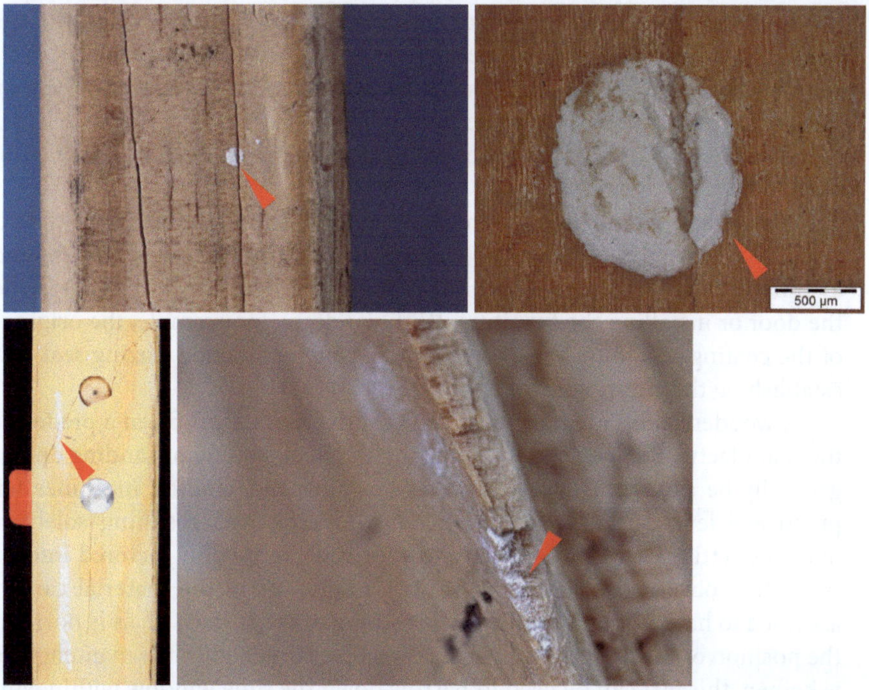

FIGURE 4.3 Top: The paint splash on the rail of a ladder obviously predates the crack: this is shown by the matching halves of the spot, which fit together perfectly when the crack faces are pressed together. Bottom: The same ladder also had a white line running across a crack. The substance was identified as gypsum. This was also found on the crack face when the crack was opened (right). The crack is therefore very likely to have existed before the mark was made.

FIGURE 4.4 The PVAc adhesive inside the reopened corner joint of a window has been stained with Lugol's iodine. The distribution of the PVAc adhesive shows that it was applied over the full face on one side (left). As the two wooden members did not fit together exactly, however, the joint faces were only in partial contact. After bonding, the white coating was able to penetrate into the rest of the joint.

by the microscopic examination of cross-sections/polished sections, it is also possible to gain decisive clues on site merely by scratching with the fingernail.

Discolored patches or scratch marks that extend across several components were obviously caused after these were joined together (Fig. 4.5). Conversely, features that end exactly at the edge of a component evidently appeared before the components were assembled.

Surface zones that were covered over at a clearly defined point in time can be particularly informative (see Sec. 4.2). These include areas below hardware, under the fair assumption that this will be fitted either prior to delivery or, at the latest, when hanging the door or installing the furniture. With windows, for example, the original condition of the coating—as found below the handle plate or silicone glazing seal—can assist in establishing the chronology of the failure.

A wooden component receives its definitive surface profile at a predefined point in the manufacturing process, that is, during the final planing or sanding operation. It can generally be assumed that modern-day planing and sanding machines will produce practically level surfaces. Moreover, any imperfections resulting solely from faulty machine settings are relatively easy to spot and can be duly factored into the analysis. Any other observed height or thickness differences in the material can therefore be assumed to have arisen after the final working of the surface. If, as is normally the case, the position of the final planing or sanding operation within the manufacturing process is known, this fact can be used to narrow down the time window within which the failure was caused. The type of information that can be deduced from this about the moisture content during production or service is discussed in Sec. 3.5.3. The example featured in Sec. 5.8 (Fig. 5.33) also shows how the thickness of a veneer helps to establish the time at which bonding failure occurred.

The original position of a surface can also be revealed by other components that directly adjoined it at the time of installation. Baseboards are a typical example: if these

FIGURE 4.5 The presence of discolored patches on two adjacent floorboards indicates that they were caused after installation.

are screwed or bonded to a wall, they will indicate the approximate level of the floor surface.

Some materials have such a characteristic surface texture that any post-production planing or sanding is immediately noticeable. Here, the presence of the original texture indicates that the material has the same thickness as at the time of manufacture. Figure 4.6 shows a crack in a continuous-pressure laminate (CPL). Microscopic examination reveals the presence of a local protrusion into the underside of the CPL directly by the crack (Fig. 4.6). The fact that this defect is almost completely filled with adhesive suggests that it existed prior to bonding. Above the defect, the white decorative paper of the CPL is approximately 30 μm thinner than over the remaining surfaces. As the surface of the CPL at this location still has the original texture, however, the defect must have originated during manufacture of the CPL.

4.1.2 Features of Bonding Failure

A piece of bread or toast with a sticky covering (honey or syrup) that is dropped and lands wrong side down on the floor will exhibit a characteristic pattern when picked up (Fig. 4.7). A similar pattern can be observed under the microscope when, for example, a bondline is pulled apart while the adhesive is still fluid (Figs. 4.8–4.10). Given that adhesives (normally) lose their fluid properties after curing, this texture on debonded

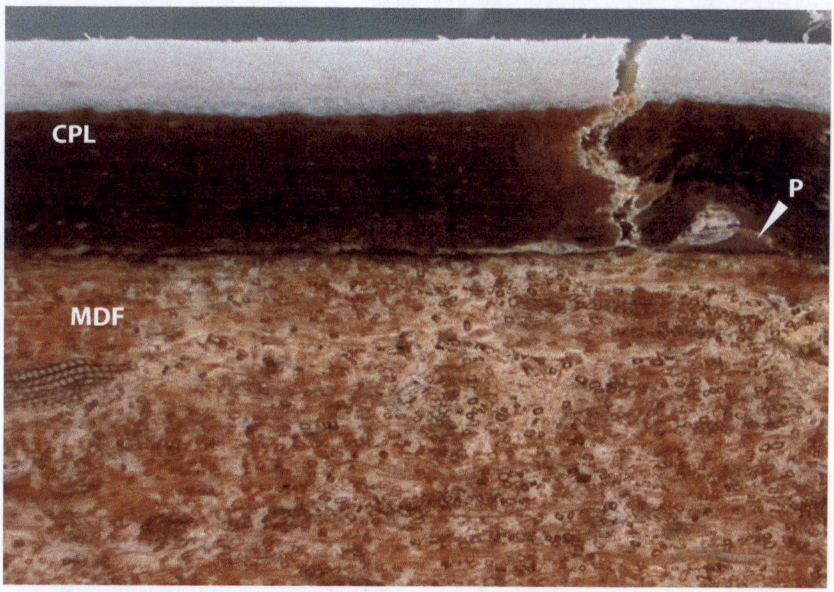

FIGURE 4.6 Polished section of a crack in a CPL, applied to a medium-density fiberboard (MDF) panel. A local protrusion (P) into the underside of the laminate can be observed next to the crack. Adhesive has entered the void created by the protrusion. Above the defect, the decorative paper of the CPL is approximately 30 μm thinner than over the remaining surfaces. The absence of any changes in the surface texture of the CPL above the protrusion indicates that the defect must have resulted from the CPL manufacturing process.

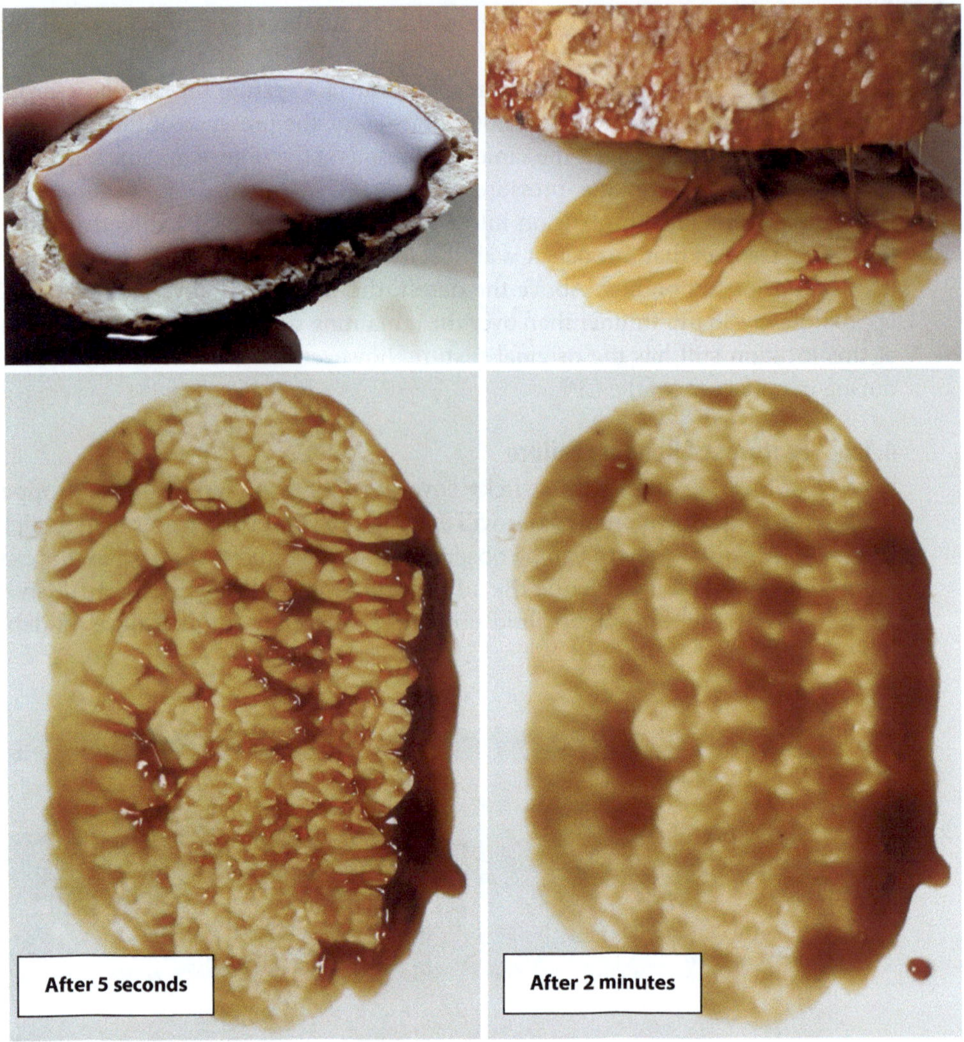

FIGURE 4.7 A slice of bread coated with syrup that lands wrong side down on the floor exhibits, when picked up, a characteristic texture that soon coalesces back to its original form.

adhesive surfaces often means that bonding failure occurred before or during curing/drying. Hence, this typical pattern—referred to here as the "honey-on-toast" texture—frequently provides a valuable clue towards understanding the chronology of bonding failures in relation to the fluidity of the deployed adhesives.

With non-thermosoftening adhesives, the honey-on-toast texture represents a highly significant finding. In such cases, any influence of the service conditions can normally be ruled out and the cause of bonding failure narrowed down to the adhesive itself or the process parameters during bonding. Such a failure may also be traceable to the adherend given that the very same finding can result where—despite the use of a suitable adhesive and pressing conditions—the adherend material inhibits the curing process.

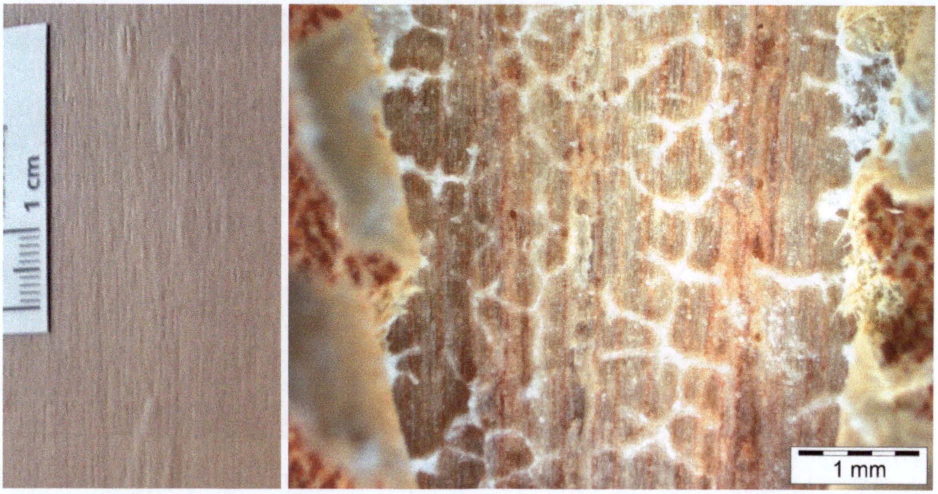

FIGURE 4.8 The urea-formaldehyde adhesive layer below blisters in a decorative paper-faced plywood panel exhibits the typical honey-on-toast texture. The blistering occurred because some areas of the adhesive had not cured sufficiently when the press was opened. Local moisture differences in the materials are one possible reason for such blisters. As soon as unambiguous findings such as this emerge, alternative theories such as those focusing on potentially unsuitable service conditions can be safely dropped.

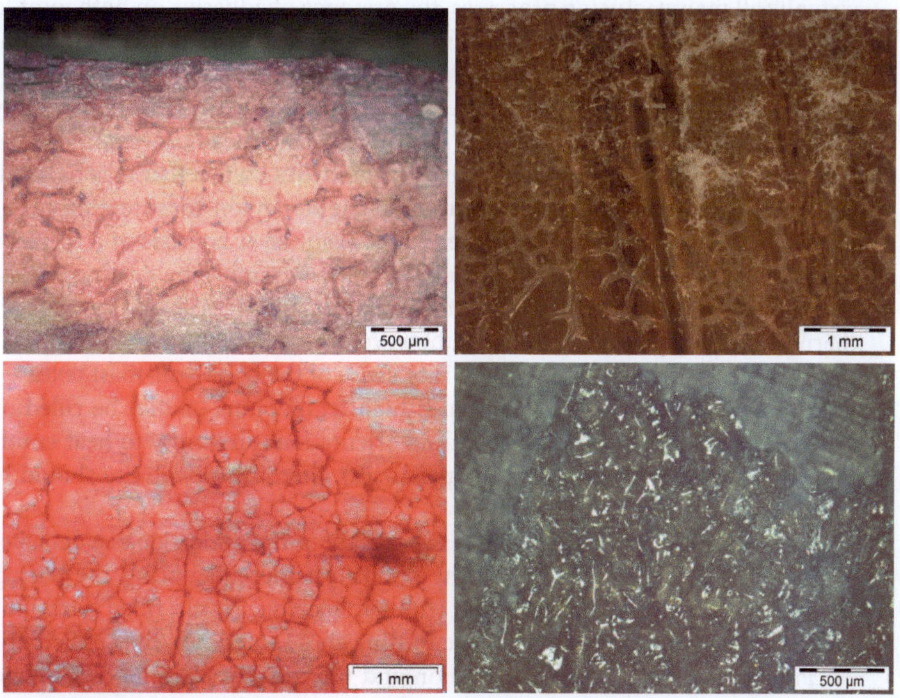

FIGURE 4.9 Various manifestations of honey-on-toast texture on adhesive surfaces.

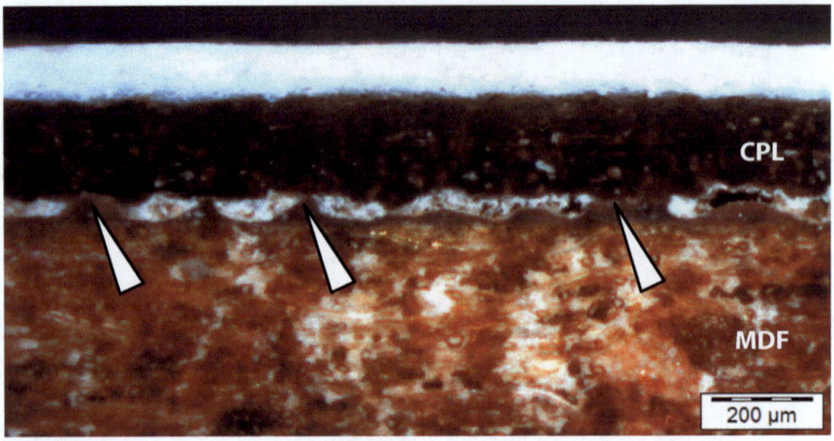

FIGURE 4.10 Polished section of honey-on-toast texture. The MDF and CPL are connected by ridges of brownish adhesive.

If the bond fails while the adhesive is still very fluid, thus allowing it to flow back together after separation of the layers, the honey-on-toast texture may disappear. In such cases, the adhesive surface will take on the same smooth and shiny appearance as adhesives that never come into contact with the opposite adherend before and during the curing process (Fig. 4.11). Given that single-side adhesive application is the norm in virtually all manufacturing processes, the presence of adhesive on both joint faces indicates that the two adherends must have been in contact at a time when the adhesive was still very fluid.

The same honey-on-toast texture is produced when thermosoftening polymer layers are pulled apart during heating (Fig. 4.12). Many thermosoftening adhesives offer adequate temperature stability only up to around 70°C, a fact that may be significant for failures in kitchen fittings. Identification of the honey-on-toast texture may therefore make it appropriate to check whether the relevant layer consists of a thermosoftening material. An indicative test can be performed, for example, by heating a metal spatula over a flame and bringing it into contact with the adhesive. The action of heat will turn thermosoftening polymers soft. It is, however, frequently possible to support or refute the hypothesis of temperature as the root cause of the honey-on-toast texture by examining the failure distribution pattern. If high temperature cannot be ruled out on the strength of failure distribution (building orientation, proximity to heat emitters, location in kitchen or under electrical appliances), then this possibility should be further investigated.

The imprint of wood on the adhesive surface or the underside of a detached coating is evidence of close contact with the opposite adherend during the curing process (Fig. 4.13). The nature of the imprint (full-surface, fine details, filled pores) may supply further information about the surface wetting and, sometimes, the applied pressure.

Figure 4.14 shows the cross-section and top view of possible findings relating to bonding failures together with the deducible facts and chronology.

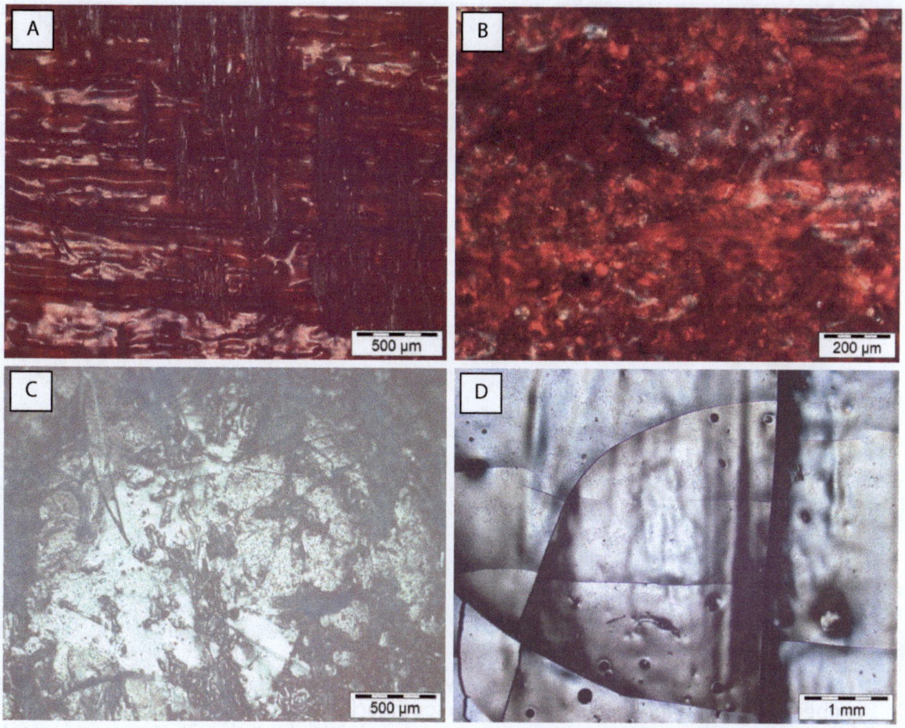

FIGURE 4.11 Examples of shiny adhesive surfaces which indicate a lack of contact with the opposite adherend during curing. A and B show PVAc adhesives stained with Lugol's iodine. Remnants of the originally bonded layer are observable in places. C and D are glint images of UF adhesives. The surface texture of adhesives is very easy to record by means of glint illumination (see Sec. 3.7.3).

4.1.3 Fungi/Insects

Due to the complexity inherent in biological systems, the time of initial infection or infestation can only rarely be estimated on the basis of the maximum growth speed (of fungi) or the life-cycle (of insects). Particularly with fungi, maximum growth speeds have often been established solely by laboratory tests and are not normally transferable to real-life situations. Moreover, the fact that long periods may also pass without any growth makes both forward and backward projections impossible.

A thorough understanding of the factors involved in degradation of a particular structure is practically hampered by a lack of detailed information on construction practices, original materials, maintenance practices, fungal species, moisture intrusion issues and a host of other factors. This type of detailed information is required to properly assess rates of decay in particular structures. In the absence of such data, investigators sometimes use combinations of laboratory and field data that may have been collected perhaps in other environments on other wood species, along with conjectures about when moisture intrusion began, to extrapolate decay rates. Often this is done erroneously. [38]

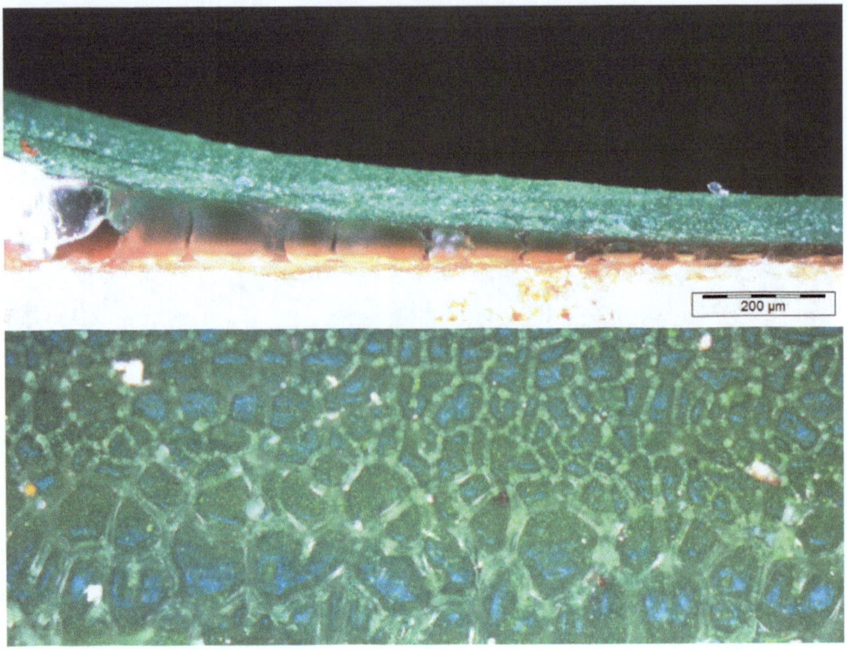

FIGURE 4.12 Cross-section (top) and top view (below) of a blister in a coating. In summer, the temperatures measured on dark coatings on south-facing elevations can reach up to 80°C. Softening of the primer upon exposure to such temperatures may lead to blistering. The honey-on-toast texture observed here confirms this as the failure cause.

FIGURE 4.13 The underside of the detached coating shows a detailed imprint of the wood surface. This suggests that the wetting was very good and the coating not overly viscous.

Cross-Section	Top View	Description and Interpretation
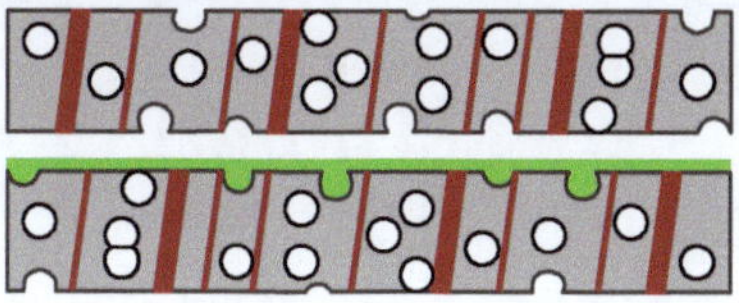		No wetting of second joint face: surfaces were never in contact.
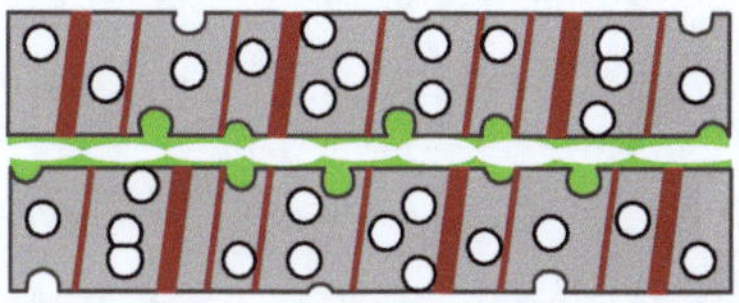		Wetting of both joint faces with honey-on-toast texture: bonding failure while adhesive was viscous. Either press opened prior to curing or, with thermosoftening polymers, excessively high temperatures during service.
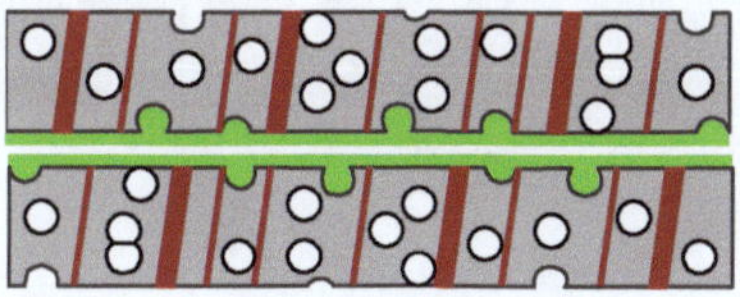		Wetting of both joint faces with shiny texture: bonding failure while adhesive was still fluid. Either press opened prior to curing or, with thermosoftening polymers, excessively high temperatures during service. Original honey-on-toast texture has disappeared due to coalescence of material.
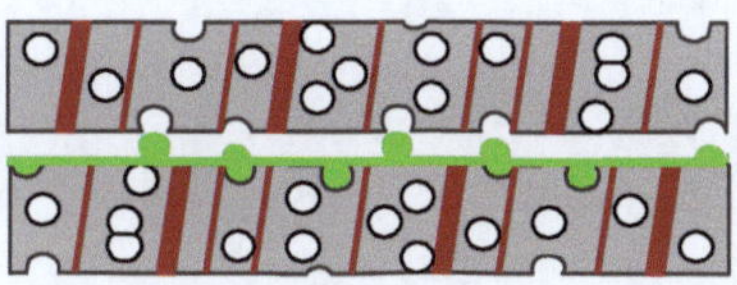		Imprint of wood surface on adhesive: failure after curing of adhesive.
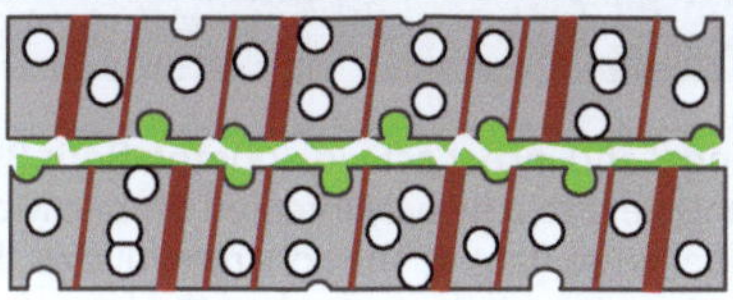	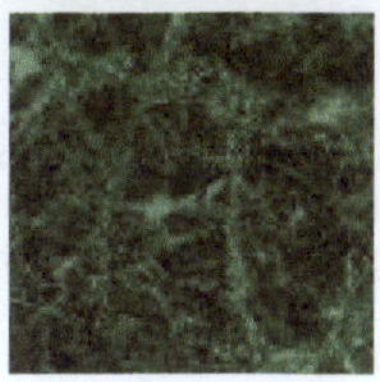	Irregular fracture within the bondline: failure after curing of adhesive.

FIGURE 4.14 Illustration of possible findings relating to bonding failure. The interpretations assume single-side application of the adhesive.

Only where the starting point and direction of fungal growth are clearly identifiable is it possible to estimate the most recent possible time of infection on the basis of the maximum growth speed. From controlled laboratory studies, it is known, for example, that the mycelia of the cellar fungus (*Coniophora puteana*) would need at least 100 days to spread a distance of one meter from its starting point [77]. Without further specific evidence, however, it would still be impossible to say anything about the earliest possible date, that is, whether the start of a one-meter-wide cellar fungus infection was 100 days or 20 years previously.

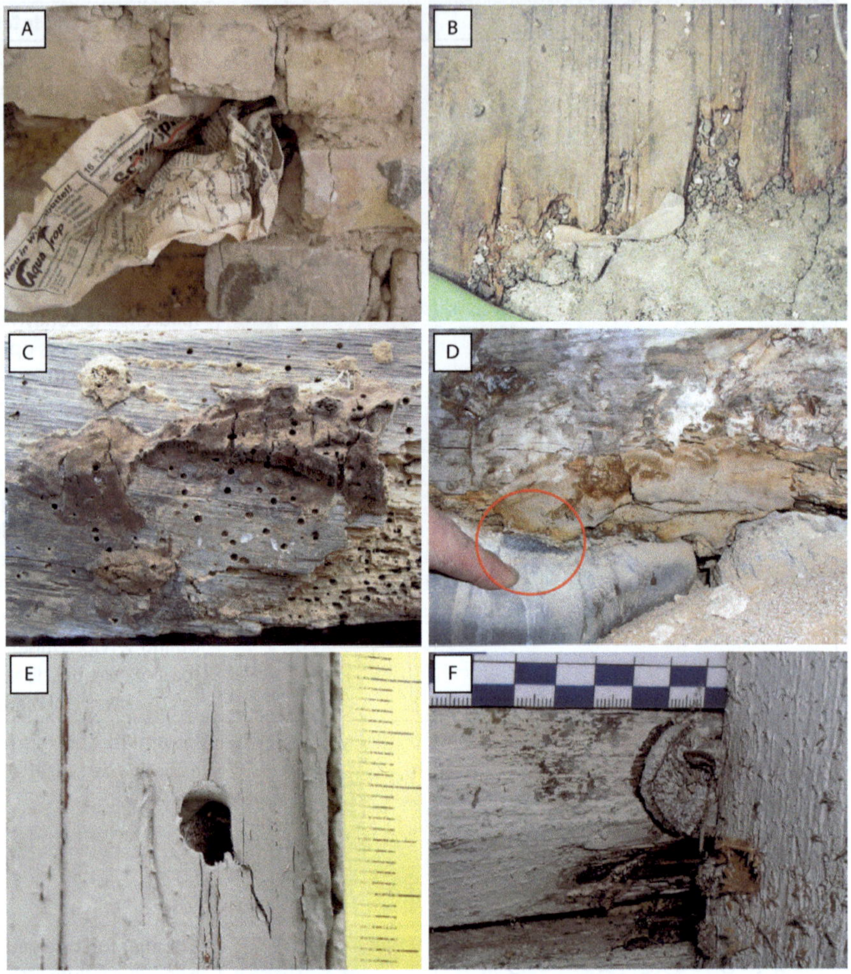

FIGURE 4.15 Hard facts providing clues as to time of fungal and insect activity (photos: Boris Blenn). A: An old newspaper without any fungal attack was found next to a heavily infected timber. Most probably, there was no fungal growth after the newspaper issue date. B: The concrete follows the geometry of the destroyed wood. By the time the concrete was cast, then, much of the fungal decay had already taken placed. C: Holes in fruiting body of a wood-destroying fungus through which larvae of deathwatch beetle emerged. The number of exit holes in the fruiting body is no lower than in the surrounding wood. The severity of the wood destruction by the insects suggests that infestation took place at least 10 years previously. The fruiting body must therefore be at least 10 years old. D: The fruiting body of a wood-destroying fungus is stuck to the insulation of a water pipe. The fruiting body (at least in its present size) must be younger than the insulation. E + F: The exit hole of a longhorn beetle and the fruiting body of a wood-destroying fungus have been overpainted. The hole and fruiting body are older than the paint finish and the painter ignored the obvious signs of biological attack.

Clues are sometimes provided by other factors offering temporal reference points: works that were obviously performed after deformations, an overcoated fruiting body, a nail driven through a fungal mycelium, subsequently incorporated organic material that is not infected. A thorough on-site search for "hard facts" of this kind can sometimes be rewarding (Fig. 4.15).

FIGURE 4.16 Piece of a door frame showing the exit holes of Lyctus powder-post beetles before and after finish is sanded away. The task was to establish whether the wood was already infected at the time of installation. With the coating sanded away, variously shaped burrows can be seen in the lower part of the specimen. Such shapes can only occur where woodworking tools have cut into existing burrows. This proves that beetle larvae were already present when the wood was worked. Additional confirmation was provided by FTIR spectroscopy analysis of material in the burrows, which was found to be wood filler.

With wood-destroying insects, important clues are often provided by the shape and condition of the exit holes. Any dirt or coating material in such holes shows that they are old or at least predate the most recent finish application.

If originally circular burrows have taken on an oval or elongated shape at the surface, then the wood must have been worked when the hole already existed (Fig. 4.16).

4.2 Distribution of Features

A consideration of the distribution of failure features allows certain causes to be ruled out immediately. Especially at the start of an analysis, this information often allows the investigator to narrow down the possibilities and concentrate on a limited number of potential causes. The distribution of features can be considered at various levels: distribution within the building, broad and localized distribution on a particular element or component, and distribution at microscopic level.

Where a failure feature is bounded by a covered area, the covering object should be removed and the exposed surface checked for the presence of the same feature. Such objects typically include furniture, pictures, and carpets, which, among other things, protect surfaces from sunlight. Even more useful, however, are fixed components such as hardware, switches, wall lamps, and baseboards. In view of the tremendous value for failure diagnosis of findings such as discolorations that either end at or extend below hardware components, it is incredible how seldom systematic investigations are performed at such locations (Fig. 4.17).

It is extremely rare for the findings below hardware components to raise more questions than they answer. As the example in Fig. 4.18 shows, the apparently plausible assumption that the discoloration below the hardware fitting rules out heat or UV radiation as a potential cause can actually be wrong.

Figure 4.17 The top images show failures not extending below hardware fittings: discoloration by a heavily dust-attracting polish (left) and fading due to UV radiation (right). The bottom images illustrate discolorations that extend below hardware fittings. The discoloration pictured on the left is due to sanding dust forced into the pores of the stain-finished composite veneer (made from the soft wood species obeche) by compressed-air gun. The defect on the right takes the form of so-called "whitening in the grain," caused by adhesive bleeding.

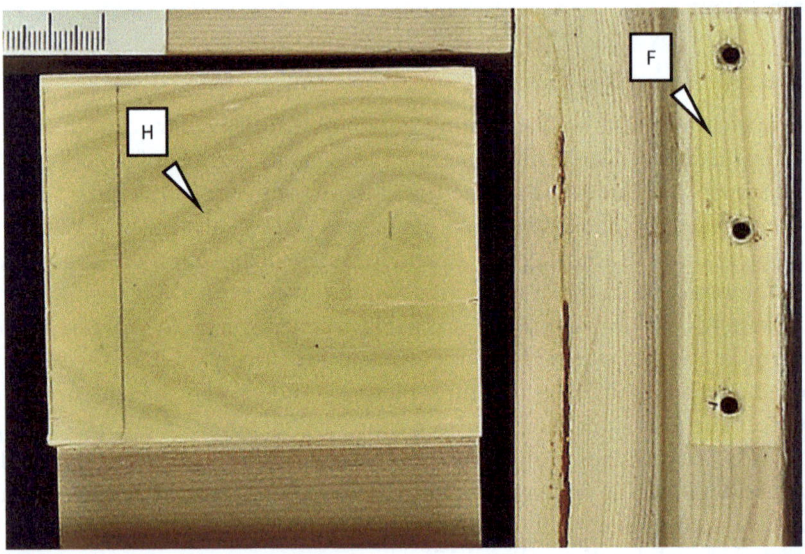

Figure 4.18 The area below a hardware fitting (F) shows the same yellow discoloration as surfaces exposed to the sun and resulting heat (H). This is one of the very rare occasions where the same features had two completely different causes. The discoloration of the transparent coating below the fitting was brought about by traces of formaldehyde from the plastics (POM) fitting, whereas the yellow discoloration in the exposed areas resulted from the heat-induced breakdown of a biocide.

4.2.1 Broad Distribution Pattern

The examples in Table 4.1, which feature a veneered wood-based panel, a finger joint, and a solid piece of wood, present possible interpretations of the wide-area failure distribution. Some of these illustrations and their interpretation are borrowed from expert reports produced by Lore and Erich Plath, the pioneers of failure analysis in Germany. While a number of examples focus on failure causes that hardly occur nowadays, they still serve to sharpen the awareness for the significance of failure distribution.

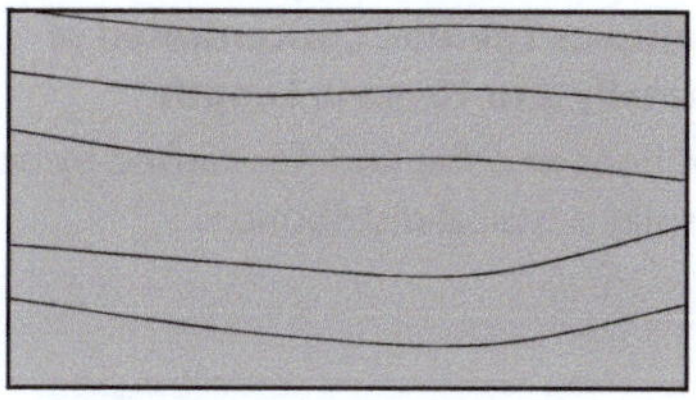	**Uniform Distribution** Unsuitable coatings, unsuitable adhesives, uniform moisture action, and exposure to gaseous substances, for example, cigarette smoke, can lead to a uniform distribution of failure features over larger surfaces.
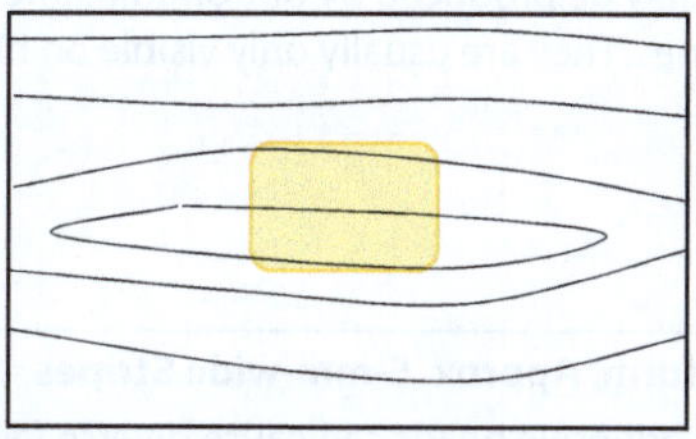	**Features in Panel Center (or Center of Original Panel)** Moisture, temperature, and steam pressure are greatest in the center of the panel during the bonding process. The bondline or film thickness may be larger or smaller in the edge zones. When panels are stacked, discoloring moisture, ammonia (from bleaching), etc., can escape more quickly from the edges. Discoloration will then tend to be limited to the panel center.
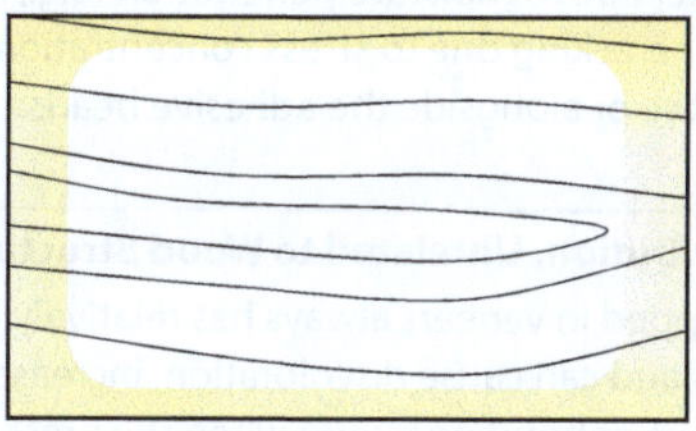	**Features Only at Edges** During hot pressing, panels may dry out more heavily in the outer zones than in the center if the temperature is too high or the press time too long. Depending on application-related factors, the bondline or film thickness may be larger or smaller in the edge zones.
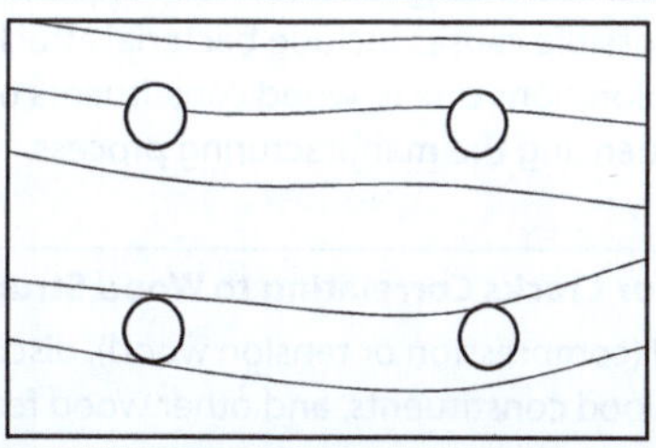	**"Systematic" Arrangement in Rows, Uniformity of Shape, etc.** Vacuum lifting devices subject veneers to loads before the adhesive has cured sufficiently. Mechanical action may result in dirtying or crushing. High-gloss films may show scuff marks left by the disk rollers of factory conveyor systems. Defects originating from applicator rolls are repeated once per revolution.

TABLE 4.1 Failure Distributions on a Veneered Wood-Based Panel and Their Possible Interpretation

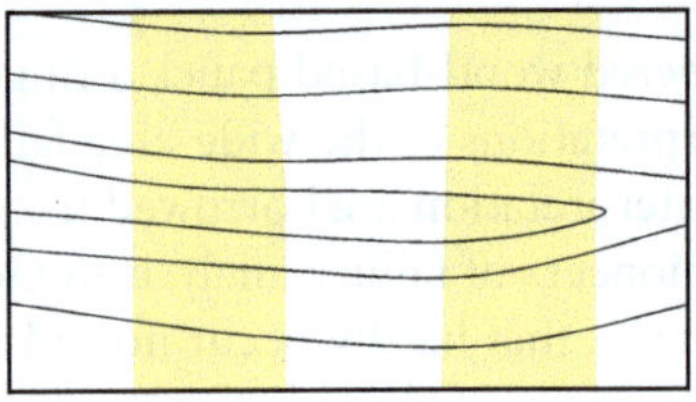	**Stripes Running Across Grain** Color extractives from cross bands may bleed through to produce stripes across the grain (sensitive veneers such as maple should only be applied to cross bands of the same light color). Similar stripes can also occur where the cross band strips were of nonuniform thickness and the thicker strips were crushed during pressing. Markings around 2 to 4 cm wide may also be caused by dirty or unsuitable stacking sticks (so-called "sticker stain"). Nonuniform adhesive or coating application is another possible explanation.
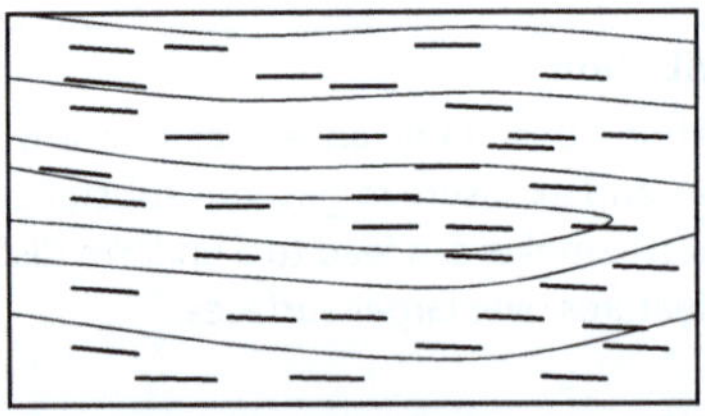	**In Grain Direction Spaced Laterally at 1- to 3-mm Intervals or Multiples Thereof. Concentrated in Latewood. Normally 2 to 10 cm in Length** Cracks, discolorations or other findings caused by open lathe checks display a typical distribution.
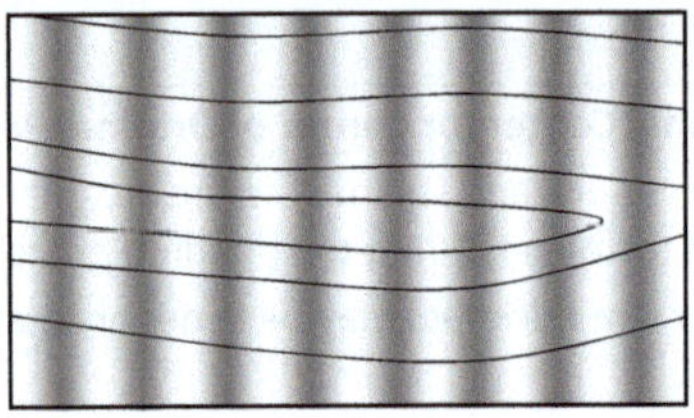	**Very Uniform Wavy Pattern (Chatter)** Chatter marks may be produced by out-of-adjustment sanding machines. They are usually only visible on high-gloss surfaces.
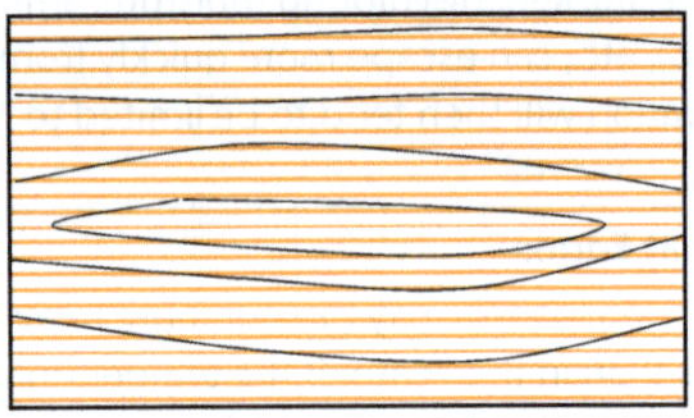	**Relatively Uniform, Approx. 5-mm-wide Stripes** Uncompressed adhesive beads can cause diverse failures, including discoloration due to adhesive bleeding, swelling of base panel due to moisture penetration (e.g., from wood stains), or cracking due to stress concentrations. The failures run above or alongside the adhesive beads.
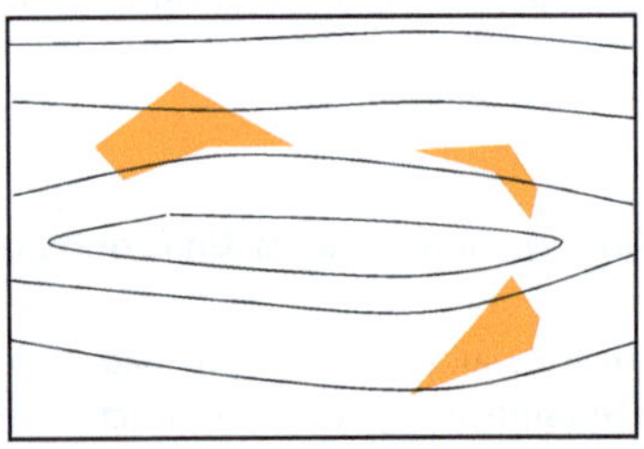	**Irregular Distribution, Unrelated to Wood Structure** Moisture entrapped in veneers always has relatively clear-cut boundaries and can cause discoloration, increased compression and/or bonding failure when the press opens. Other possible causes include bacterial attack, abnormal discolorations due to wood constituents or various factors influencing the manufacturing process.
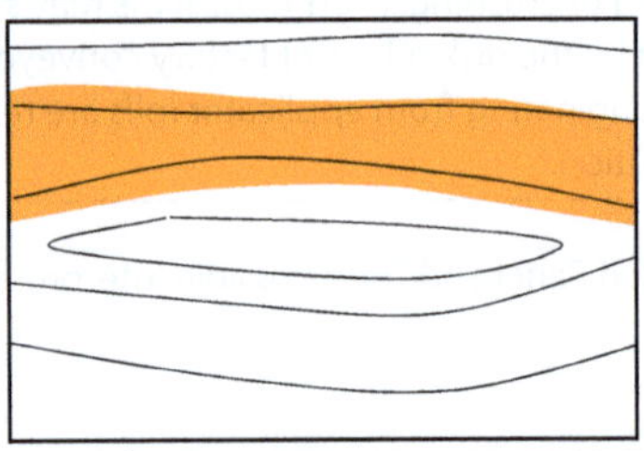	**Discoloration or Cracks Correlating to Wood Structure** Reaction wood (compression or tension wood), discolorations due to wood constituents, and other wood features tend to echo the wood structure.

Table 4.1 Failure Distributions on a Veneered Wood-Based Panel and Their Possible Interpretation (*Continued*)

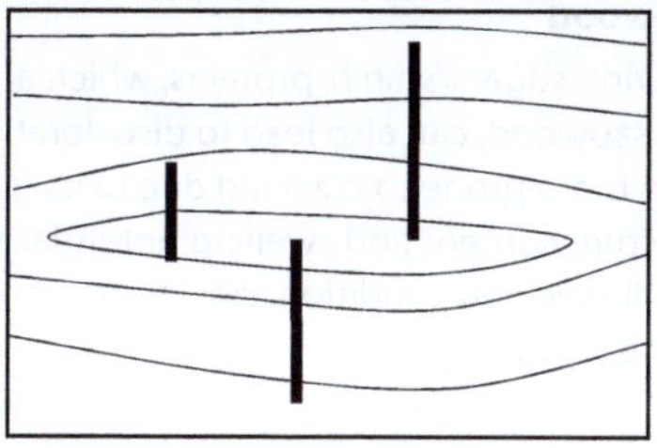	**Discoloration or Unevenness Running Across Grain** Lathe checks in the cross band may have opened and been filled with adhesive during bonding of the face veneer. The resulting local differences in moisture or pH value may then bring about the discoloration of sensitive face veneers. Where individual strips in the core layer of blockboard are excessively dry or moist during manufacture, these may show through during service (telegraphing).
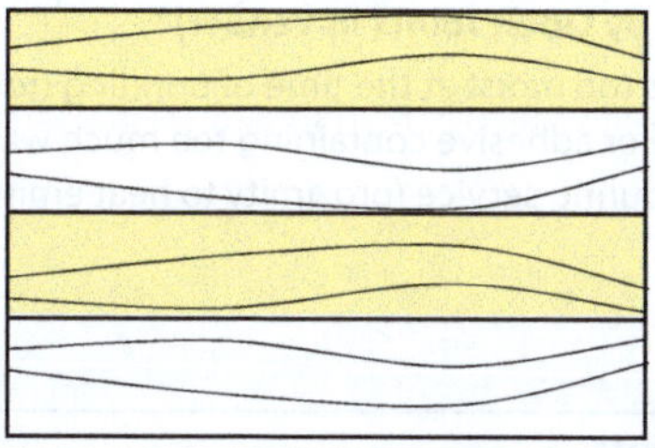	**Every Second Strip of Book-Matched Veneer** When oak dries, tannins tend to migrate to the so-called "loose side" (with the surface checks) where evaporation is faster due to the larger surface area. The initially yellow tannins subsequently darken due to oxidation. With book-matched veneer, the loose side alternately faces upwards and downwards.
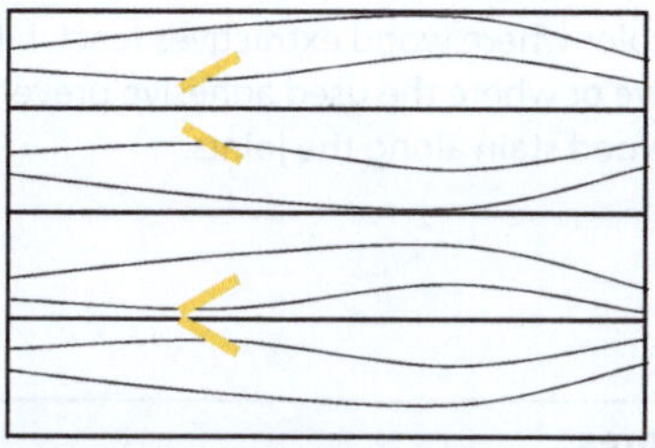	**Every Strip of Book-Matched Veneer (Not at Exactly Matching Position)** Although wood features such as discolorations, slope of grain and knots recur on the individual strips of book-matched veneer, their position and size gradually change.
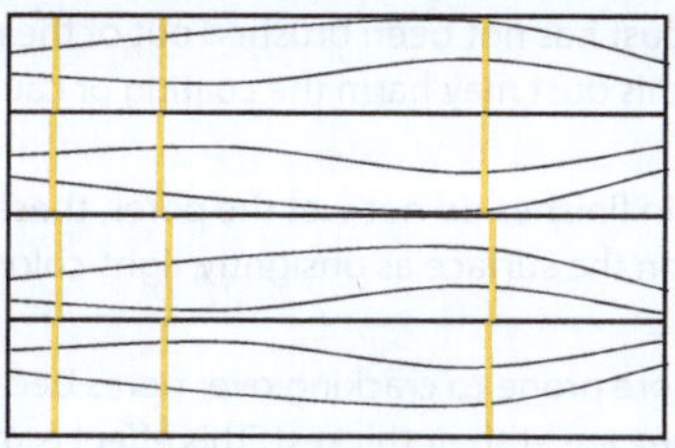	**0.5- to 2-mm-wide Stripes, Exactly Perpendicular to Grain (Not at Exactly Matching Position)** In veneer production, larger chips in the knife edge will cause the localized crushing of wood fibers. This may result in the increased or reduced absorption of adhesive, stain finishes or other coatings in the relevant zones.
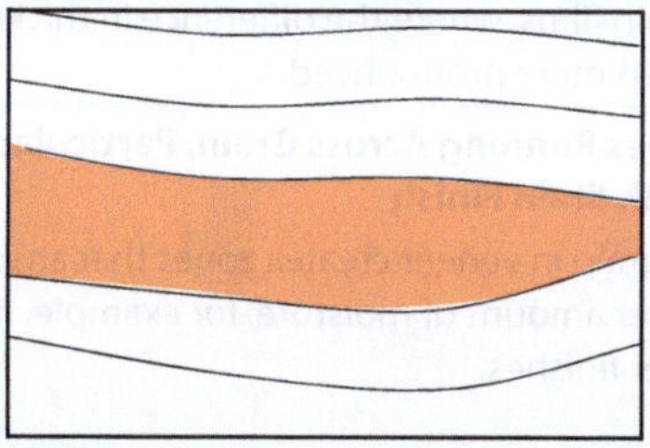	**Only Over Heartwood** Steam-volatile softwood resins are driven to the surface during pressing. Resins can cause coating adhesion problems. Heartwood extractives can retard the drying of coatings or produce gray discolorations.

Table 4.1 Failure Distributions on a Veneered Wood-Based Panel and Their Possible Interpretation (*Continued*)

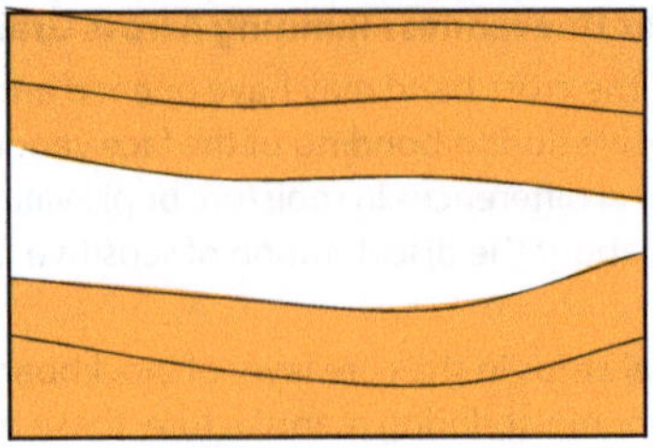	**Only Over Sapwood** Reactions involving sugars/starch/proteins, which are only present in sapwood, can also lead to discoloration. Sapwood is also more prone to cracking due to its generally higher moisture content and swelling potential. If it is bonded in heavily swollen condition, cracks are more likely develop upon heavy drying.
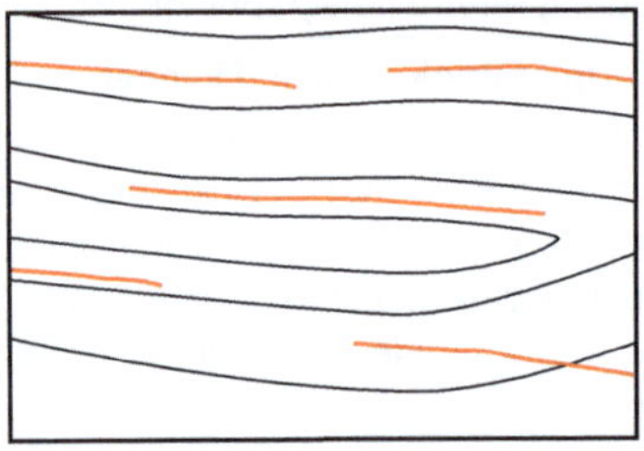	**Typically Long Cracks Exactly Parallel to Grain (Often Accompanied by Open Joints in Veneer)** The veneer was too moist at the time of bonding (too much adhesive or adhesive containing too much water) or has overdried during service (proximity to heat emitter).
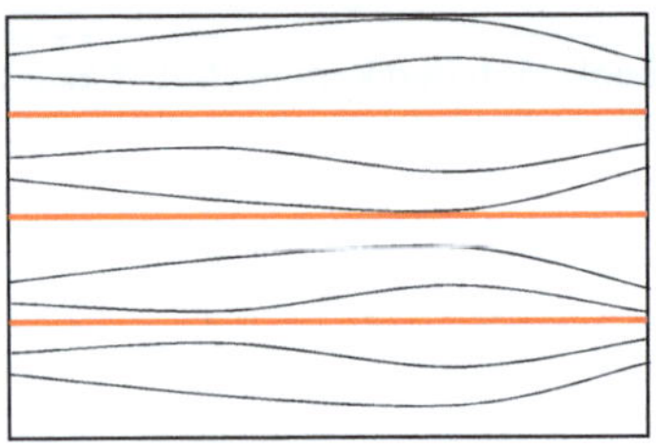	**Discolored Joints** Joints may discolor where wood extractives react with the jointing adhesive or where the used adhesive prevents the absorption of wood stain along the joints.
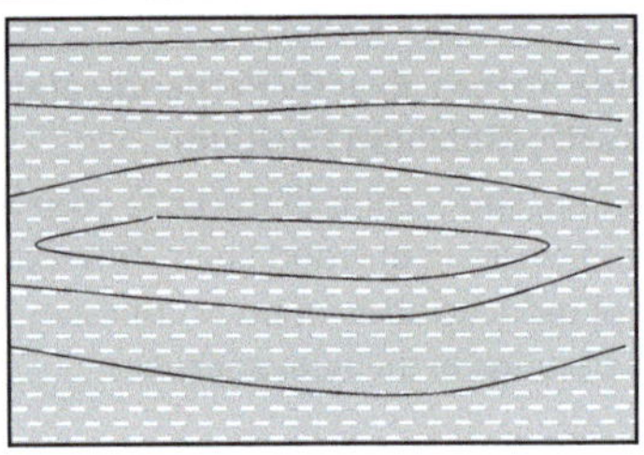	**Over Larger Pores** If the sanding dust has not been brushed out of the pores, air rising from this dust may harm the coating or cause discoloration. Where dark stain finishes do not wet the pores, these will remain visible on the surface as unsightly, light-colored marks. Coatings are more prone to cracking over pores because this is where they are at their thickest. This effect is most evident with thin films, where the difference in thickness at the pores is even more pronounced.
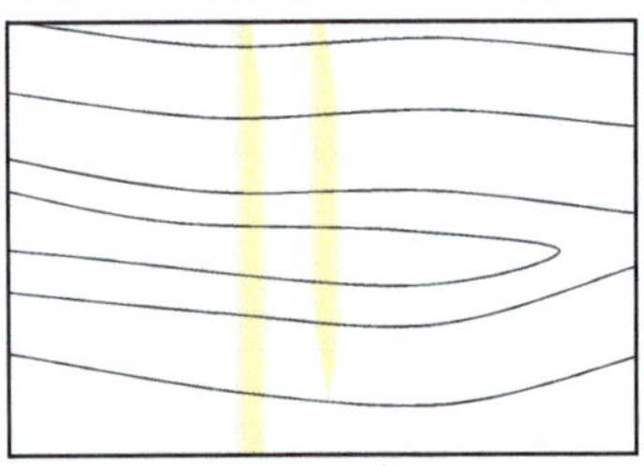	**Irregular Stripes Running Across Grain, Particularly Noticeable with Stain Finish** Local slope of grain in veneer creates zones that absorb a higher or lower amount of moisture, for example, from stains and other finishes.

TABLE 4.1 Failure Distributions on a Veneered Wood-Based Panel and Their Possible Interpretation (*Continued*)

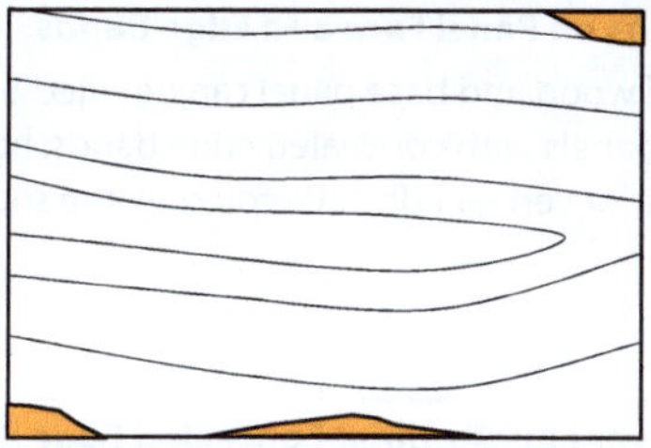	**Bonding Failure at Panel Edges** A poor bond that affects the entire surface tends to manifest itself first at the panel edges.
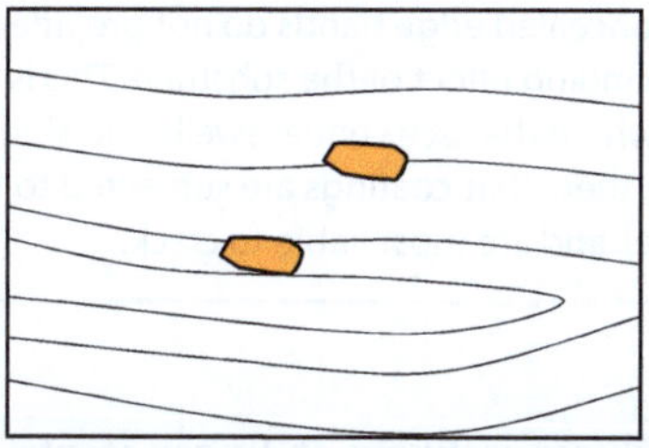	**Bonding Failure in Panel Center** Blisters or wide surface cracks in the panel center suggest that a local phenomenon is the cause of the failure and that the bond is not defective over the entire surface. Otherwise, bonding failure would first be expected at the edges.
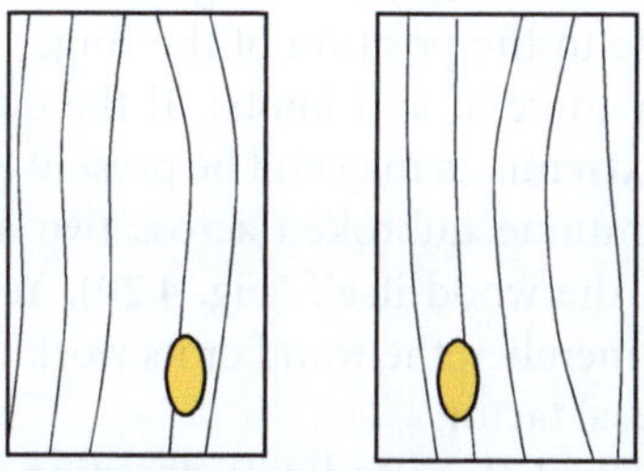	**Matching Feature on Rear Panel Face** Oil or other contamination on press platens that come into contact with each other will produce similarly shaped patches on the upper and lower panel faces.

Table 4.1 Failure Distributions on a Veneered Wood-Based Panel and Their Possible Interpretation (*Continued*)

In the case of panels with concealed or non-concealed edge bands (i.e., edge bands bonded before or after the face veneers), the interpretation can also take account of any differences in failure distribution between faces and edges.

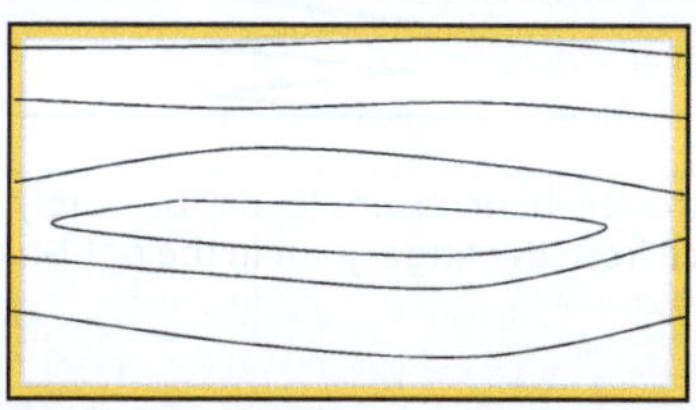	**Failures Only Over Edge Bands** The edge band wood must be the cause of the problem.
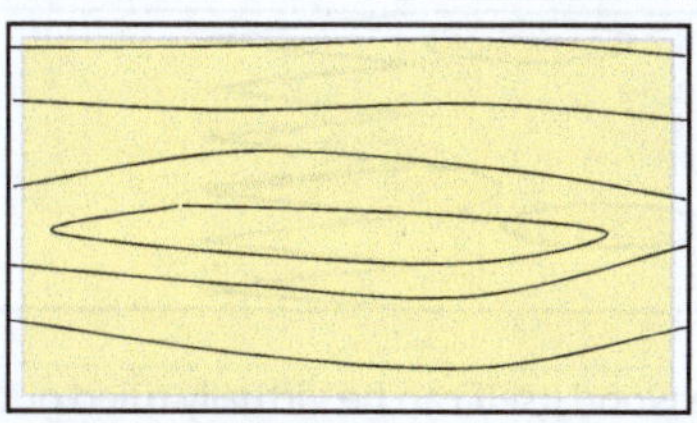	**Failures Only on Panel** The panel must be the cause of the problem. For panels with (unaffected) non-concealed edge bands, the veneer or veneer adhesive could also be the cause.

(*Continued*)

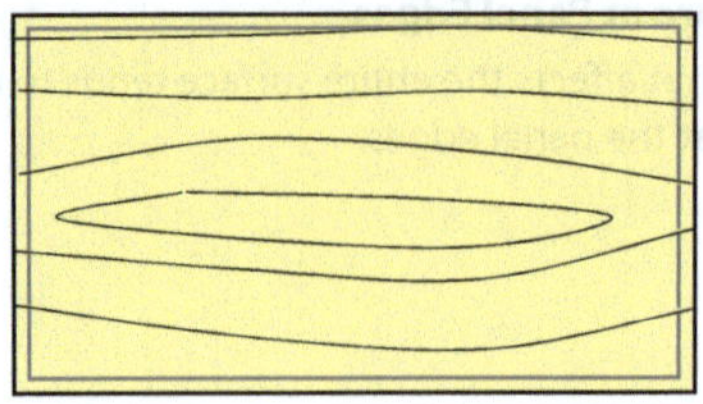	**Similar Findings on Panel Face and Edge Bands** The edge band wood and base panel can be ruled out as the cause. For panels with concealed edge bands, however, the veneer or veneer adhesive could be the source of the problem.
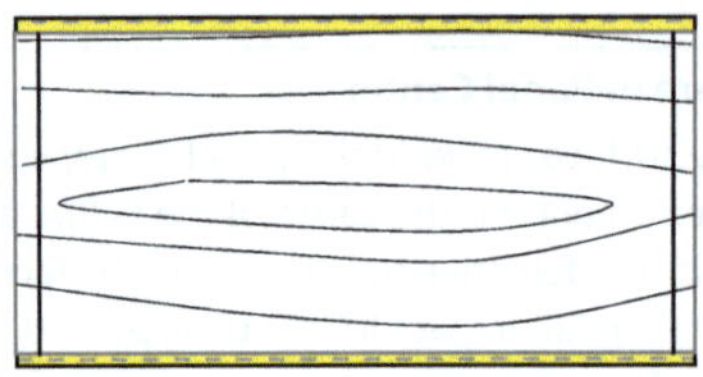	**Cracks Only Over Longitudinal Concealed Edge Bands** Longitudinal concealed edge bands do not provide the normal cross-banding effect of the substrate. This is consequently where the face veneer swells and shrinks the most. It is here, then, that coatings are subjected to the greatest stresses and are most liable to crack.

An increased incidence of wood-related failures is likely at finger joints. This is because wood members are often trimmed close to the position of the finger joint to remove undesirable features (cracks, high resin content, and knots). If the cut is not made far enough away from such a feature, its extremities may still be present near the finger joint (Fig. 4.19). Failure features that continue unbroken across two adjacent wood members are very unlikely to stem from the wood itself (Fig. 4.20). Yet, if the features are found only at the end of one mating member, the wood or its working prior to bonding/assembly can be assumed to be a causal factor.

Wood absorbs liquids far more rapidly in axial (longitudinal) direction than in transverse direction (through the side faces). At the same time, many production processes involve making cuts across the length of timbers at a fair distance from the original surface while, if at all, only a few millimeters are shaved off the side faces. The three

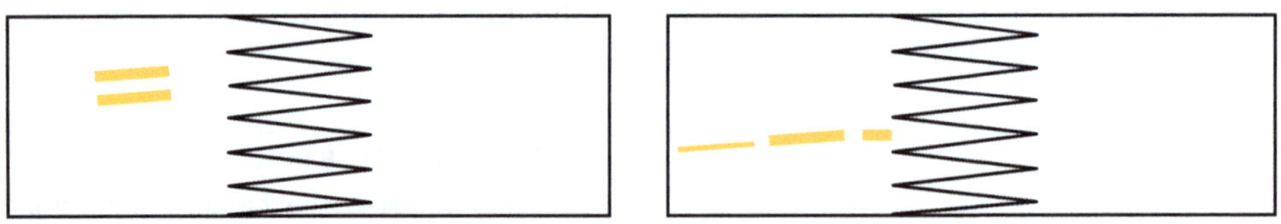

FIGURE 4.19 While the position of the failure features in the left-hand example may be purely accidental, their higher incidence with decreasing distance from the finger joint in the right-hand example makes inadequate trimming the likely cause here.

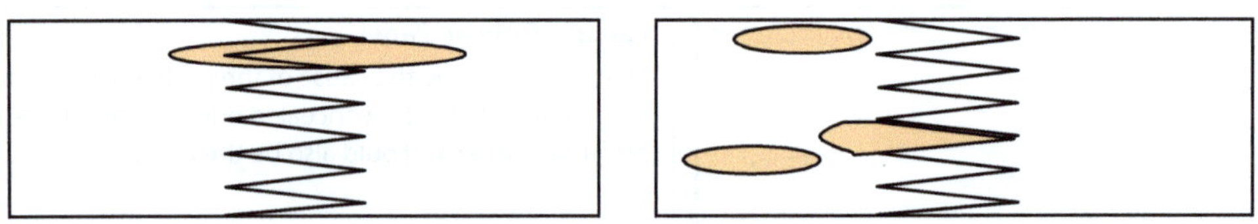

FIGURE 4.20 Failure distribution at finger joints. Left: The wood itself can be virtually ruled out as the cause. Right: The wood is very likely to have been a causal factor.

alternatives for the distribution of discoloration over side and end grain pictured in Fig. 4.21 would—assuming an otherwise uniform treatment and loading of the timbers—permit possible interpretations.

4.2.2 Distribution in and on Building

The orientation of a building has direct implications for its exposure to rain and UV as well as the temperature range on the facade surface. It also affects wind loads and the resulting pressure distribution inside the building.

- In central Europe, dark, south-facing surfaces can reach temperatures of up to 80°C. By contrast, the same surface on northerly elevations can only reach the maximum air temperature of approximately 40°C. Nor are north-facing building compliments likely to suffer any notable damage from UV radiation in Europe and North America.
- Condensation problems due to leaks in the building interior tend to occur on the leeward—that is, normally eastern—building front. (However, consideration should also be given to local conditions and temporary variations in wind direction.) On the windward side, outside air can infiltrate into buildings, for example, through unsealed windows. The resulting positive pressure inside the building pushes out air through leaks in the windows on the opposite side and may cause condensation problems on this facade.
- Humid air is lighter than dry air and warm air is lighter than cold air. As a rule, then, the absolute water content of air will rise in proportion to the height within

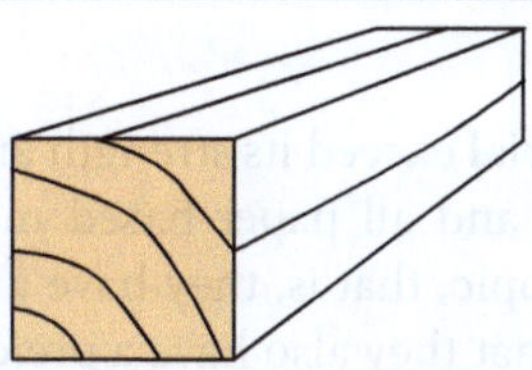	Side faces, though not end grain, have been planed, sanded, etc. Over-absorption of impregnation, stain, gas, etc., by end grain.
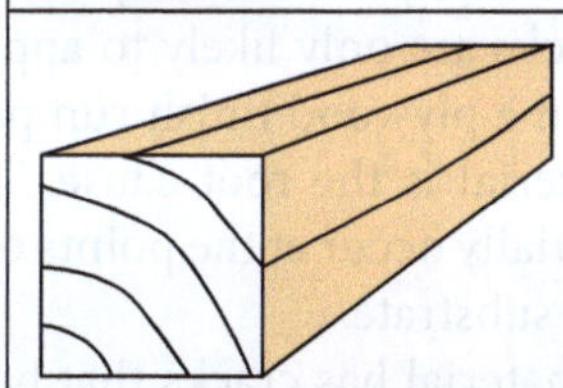	Discoloration appeared prior to final trimming.
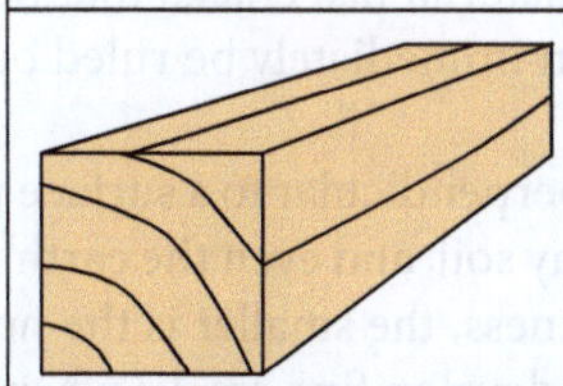	Either cause of discoloration inherent in all parts of wood or external cause after final trimming.

FIGURE 4.21 Possible distribution of discoloration over side and end grain.

the building or space. Humidity-related failures are thus more likely to occur on the upper stories of a building.

- Projecting features will protect underlying facade surfaces against driving rain (assumed angle of incidence: 30° to the vertical).
- Splash water can be expected to reach a height of 30 cm above the ground level. Even without exposure to splash water, however, the quantity of water acting on lower facade areas after heavy rainfall will often exceed the absorption capacity of the wood. In such cases, any splash water exposure cannot increase the quantity of water absorbed by the wood.
- Regarding the distribution of failures inside of buildings, factors to be considered include the proximity to heat emitters and all relevant aspects of user behavior (cleaning, wiping, mechanical action).

4.3 Interpretation of Cracks and Checks

An examination of the shape and distribution of cracks and checks, even in the early stages of an analysis, will point to a variety of potential causes while largely ruling out others. Possible questions that arise in the course of a failure analysis may focus on whether the cracks were initiated by:

- Properties of the coating
- Properties of the substrate
- Properties of the adhesive bond
- Climatic or environmental factors
- A combination of effects

Cracking results where the stresses acting on a material exceed its strength and cannot be relieved by deformation. Solid wood, veneers, and all paper-based materials (decorative papers, high-pressure laminate) are anisotropic, that is, they have a preferential direction with the greatest strength. This means that they also have a preferential direction for cracking. With plywood, for instance, cracks are only likely to appear in the grain direction of the face veneer. Hence, if cracks in a plywood finish run parallel to the grain, it is not immediately obvious which material is the root cause. This is because cracks caused by a defective coating will also initially occur at the points of high stress concentration as dictated by the anisotropy of the substrate.

On the other hand, if a coating on an anisotropic material has cracks that bear no relation to the substrate structure, the substrate itself can immediately be ruled out as a principal cause of cracking (Fig. 4.22).

The mechanisms governing the formation of cracks perpendicular to a surface would appear to be similar for drying coffee powder, parched clay soil, and even the earth's crust [37]. As a rule, the greater the crack depth or layer thickness, the smaller is the number of cracks [25]. For example, while thin layers of clay soil develop fine cracks when dried out by sunshine, deeper layers will break up into plate-sized chunks. At the same time, the width of the cracks increases in proportion to the depth of their origin (Fig. 4.23).

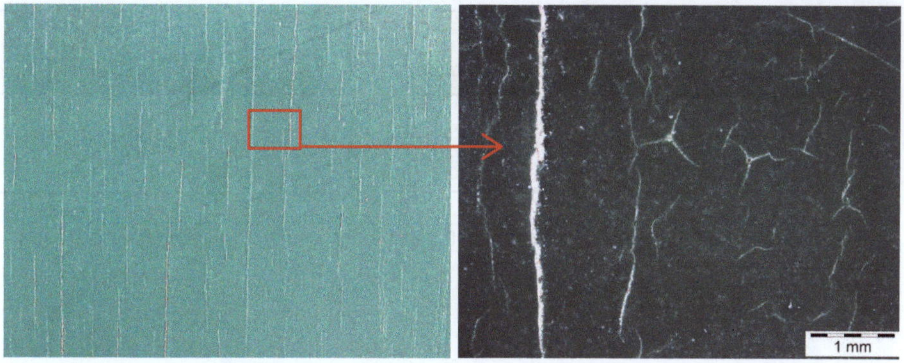

FIGURE 4.22 Large cracks, visible with the naked eye, in the paint finish of a door run parallel to the grain of the face veneer. Cracks of this kind may have been caused either by the coating or the substrate. However, examination under a magnifying glass reveals the presence of numerous small cracks that do not run in the preferential direction of the substrate. This strongly suggests that the coating is to blame. Possible causes include the choice of an unsuitable paint finish, application errors (e.g., at excessively low temperatures) or faulty batches of material.

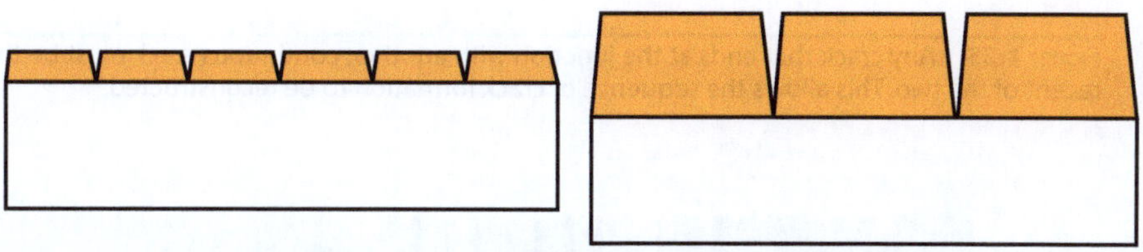

FIGURE 4.23 The deeper the layer, the smaller the number of cracks and the greater the crack width.

Cracks in a layer that is firmly attached to the substrate will exhibit a V-shaped cross-section. Accordingly, the resistance to this V-shaped deformation of the cracking material will determine the particular crack features. In hard materials with high deformation resistance, the cracks will remain narrow. Soft and thus more easily deformable materials, on the other hand, can exhibit much wider cracks. Given that the degree of stress relief in materials increases in proportion to the crack width, harder materials will—assuming identical brittleness and stress levels—exhibit a larger number of smaller cracks compared to softer materials (Fig. 4.24).

A particularity of cracks is that they almost always converge with neighboring cracks at a roughly 90° angle. This allows the chronology of crack formation to be reconstructed (Fig. 4.25). Any crack that ends at the junction with another, continuous crack must be the more recent of the two. It is also a fair assumption that long continuous cracks will always predate smaller cracks.

Crack propagation is encouraged by inhomogeneities and local weaknesses. Homogeneous layers are much less prone to cracking than layers which inherit many potential crack starting points [25]. A large number of fine cracks indicates a physically weak layer, whereas fewer cracks suggest that the layer strength is adequate and that other factors, for example, swelling, are responsible for the failure (Figs. 4.26 and 4.27).

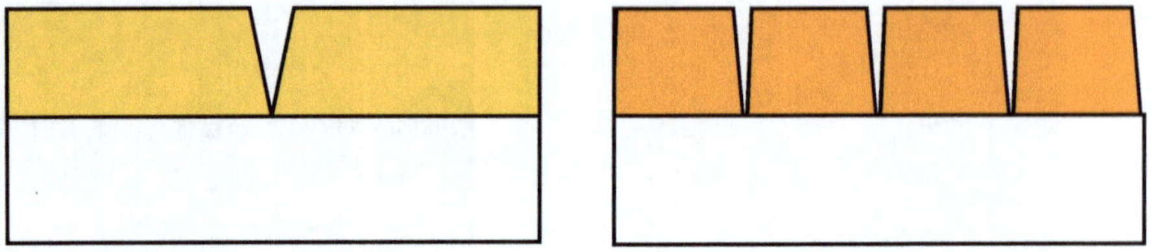

Figure 4.24 Soft materials tend to exhibit fewer, though wider cracks than hard materials.

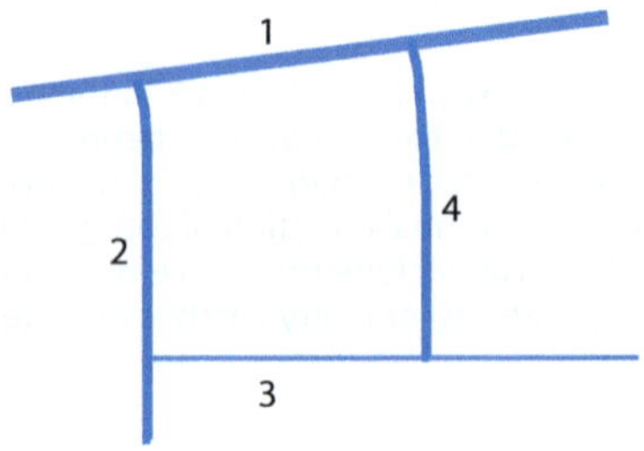

Figure 4.25 Any crack that ends at the junction with another, continuous crack must be the more recent of the two. This allows the sequence of crack formation to be reconstructed.

Figure 4.26 Crazing in a finishing coat without the typical convergence angle of 90° between cracks. Although the priming coat did not fully wet the pores, it is free from cracks. The shape of the cracks in the finishing coat indicates a basic weakness in the coating material. The probable cause was either a faulty material batch or frost damage to the coating while it was liquid or before it had fully cured.

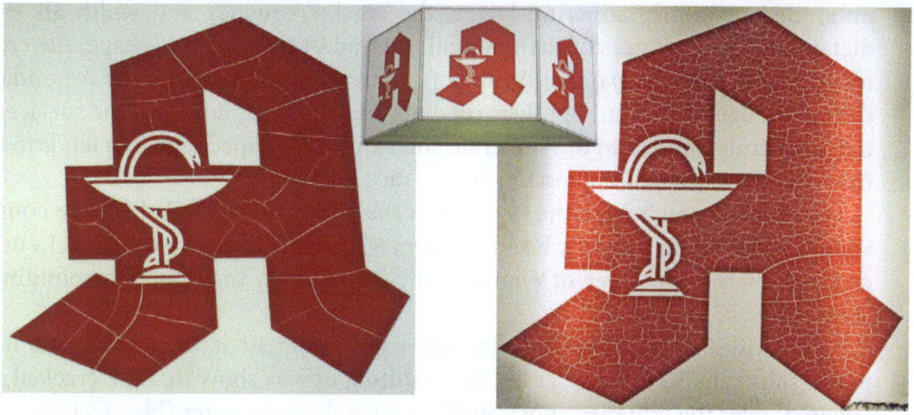

FIGURE 4.27 The foil letters of a pharmacy sign exhibit varying crack patterns depending on the building orientation. The side hardly exposed to sunshine (left) shows normal shrinkage cracks that typically originate from notches and converge at an angle of 90°. On the side heavily exposed to sunshine (right), UV radiation damaged the film so severely as to produce a crazed pattern, with cracks converging at various angles.

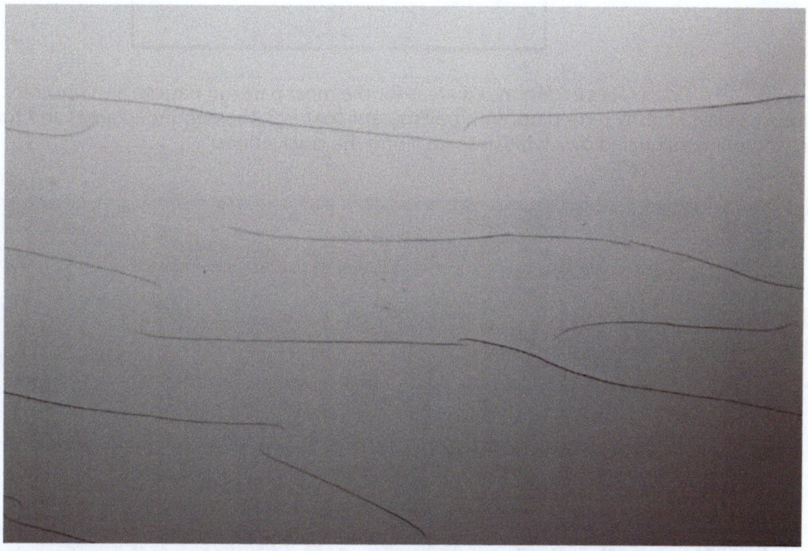

FIGURE 4.28 Even though the cracks exhibit a preferential direction, the gently undulating crack line and sometimes clearly curved tails suggest that the coating is responsible for the failure.

Cracks that are caused by the coating itself typically have a curved tail. Indeed—even where the cracks tend to run in the preferential direction of the substrate—curved tails in longer cracks provide an initial indication that the coating is to blame (Fig. 4.28).

Layers that are not attached to the substrate are subject to completely different stress conditions from layers that are firmly bonded. This difference shows itself in the cracking behavior, specifically in the frequency and width of the resulting cracks. With unbonded or poorly adhering layers, the crack frequency and width are no longer dictated by the hardness of the material. Instead of forming a V-shape, the crack faces for the most part run parallel. With soft materials and only partially debonded layers, the crack edges are often slightly turned up (Fig. 4.29). By contrast, the surface of layers that are firmly bonded to the substrate and exhibit V-shaped cracking tends to be flat or may even have a slight depression at the crack edges.

Cracks in areas where the bond with the substrate has failed can be considerably wider than those in properly bonded layers. Consequently, very wide cracks in veneers, coatings or the top layer of engineered parquet, etc., may point to bonding failure (Figs. 4.30–4.32).

Very wide cracks in thin films indicate a displacement of the cracked layer in relation to the substrate. In extreme cases, the resulting offsets show that the cracked layer has "floated" over the liquefied substrate like an ice sheet on water (Fig. 4.33).

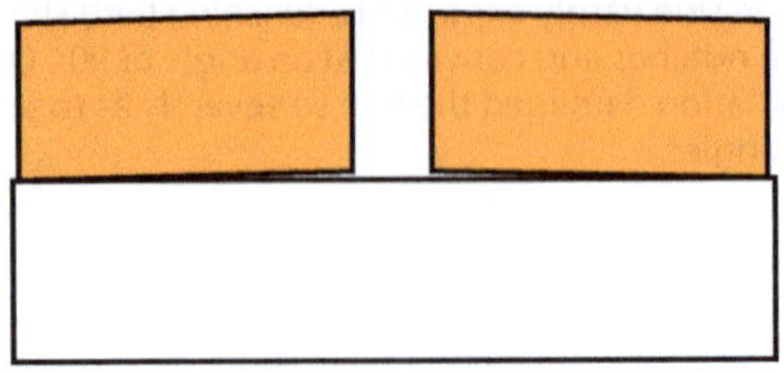

FIGURE 4.29 Crack faces in debonded areas for the most part run parallel and the crack edges are almost always slightly turned up. With transparent coatings, the debonded areas and turned-up edges are sometimes signaled by a whitish seam along the crack edges.

FIGURE 4.30 Very wide cracks on the surface of composite materials are frequently a symptom of adhesion loss. Bonding failure allows the formation of a wide crack with parallel faces (left). Where the bond is (still) intact, the crack will be V-shaped and of accordingly limited width (right).

4.3.1 Cracks in Edges of Medium-Density Fiberboard

The coated edges of medium-density fiberboard (MDF) are sometimes susceptible to cracking. Unlike the faces, the edges are not covered with primer paper. Also, with regard to the panel geometry, they lie in the plane of the lowest cleavage strength, highest water absorption, and greatest swelling. Past experience has shown that the causes of

FIGURE 4.31 Micrograph showing thin section of coating on pine. On the right, a V-shaped crack can be seen in the coating. The parallel-face crack (left) passes through microfoam bubbles at a location where the bond between coating and substrate has failed (arrows).

FIGURE 4.32 Cross-section through beech plywood. Wide cracks in the face veneer that occur only in the panel center (left) and not at the edges are a sign of poor adhesion of the face veneer. This is manifested at the panel edge by bonding failure without cracking.

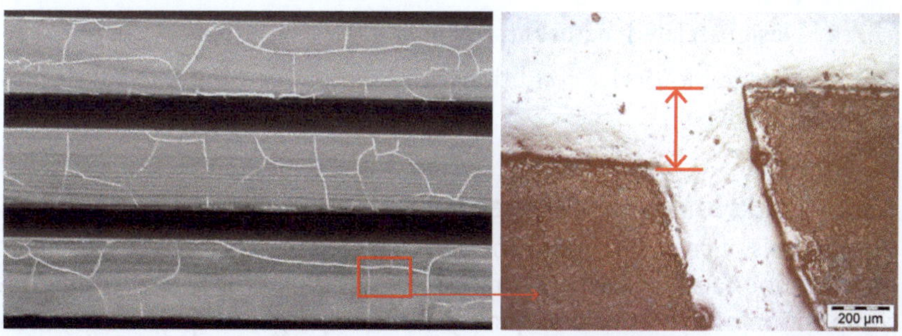

FIGURE 4.33 Wide cracks in a coating that do not reflect the wood structure can only form where there is no proper bond with the substrate. In this particular case, the fireproofing primer had liquefied due to moisture action. This caused the thin finishing coat to break up like floating ice. The offset in the "floating" film, observable under the microscope, confirms the prior liquefaction of the priming coat.

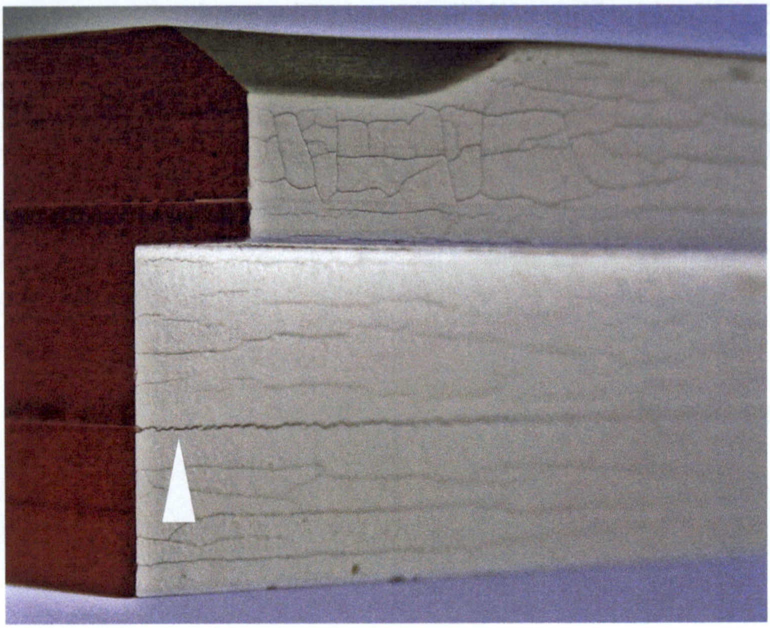

FIGURE 4.34 The pattern of the cracks indicates that they were caused by the coating material or application. A continuous crack can be seen to run along the bondline between two panels (arrow), though this is only to be expected with crack-prone coatings.

cracks in the edges of MDF cannot always be clearly established. Such cracks are obviously induced by a variety of factors:

- Absent, sanded-away or defective filler and priming coats
- Overly thick coatings or strong local variations in coat thickness
- Unsuitable finishing coats (Fig. 4.34)
- MDF with inadequate internal bond strength and abnormal density profile (Fig. 4.35)
- Design-related features (see Fig. 4.47)
- Unsuitable service conditions, for example, causing extreme dryness or moisture in panels

4.3.2 Cracks in Melamine-Impregnated Papers

The analysis of cracks in melamine-impregnated papers (HPL, CPL, paper foils, facings of melamine-faced boards, decorative overlays) is often a challenging task. The exact composition of the resins and quality of the impregnated papers may show wide variations depending on the manufacturer. Nor do the official standards covering these materials offer much help. Moreover, the resins may undergo changes over time. This makes it even more difficult to distinguish between the individual effects of the material, application process, substrate, and service conditions. The pattern of cracks may provide initial clues as to their cause and suggest methods for further analysis (Table 4.2).

FIGURE 4.35 These cracks run across a zone where the MDF exhibits its lowest strengths. The crack shape suggests that inadequate cleavage strength may be one of the contributing factors to the failure. In this particular case, the panel's internal bond strength was very low, falling far short of the values specified in the relevant standard.

Crack Type	Cause	Method for Further Investigation	Comments
Long unidirectional cracks in production direction of HPL/CPL or at angle up to 15°. Often two or more parallel cracks (Fig. 4.36)	Inhomogeneous bond with substrate	Microscopy, sanding down, taper sanding	
	Extremely dry air	Check surrounding materials for signs of shrinkage	Especially on non-flammable boards without shrinkage: the stiffer the adhesive, the shorter the distance between parallel cracks
	Cracking resistance of material too low/ material shrinkage too high	Determination of shrinkage (e.g., 24 h/ 70°C), standard test or enhanced standard tests for cracking resistance	Requirements of standard tests (EN and AS/NZS) not very stringent
	Papers swells before adequate adhesion achieved	Cutting free of shrinkage stresses, microscopy (adhesive thickness), IR spectroscopy (paper without resin)	Particularly common in paper assemblies where some layers are without resin
Originating from corners of cut-outs. Often starting with curve, then unidirectional (Fig. 4.37)	Stress concentration	Ensure that other factors can be ruled out	Regardless of root cause, first cracks will always appear next to areas of stress concentration

TABLE 4.2 Possible Interpretations of Crack Patterns in Melamine Surfaces (HPL, LPL, etc.) and Methods for Further Investigation

Crack Type	Cause	Method for Further Investigation	Comments
Various patterns limited to single area	Local heat source (e.g., hot spots, printers)	Distribution	
Irregular cracks in surface paper (Fig. 4.38)	Decorative paper with insufficient amount of resin	Check whether splitting with thumbnail is possible, uptake of dyed water, determination of resin content, for example, by IR spectroscopy	
Various small patterns (Fig. 4.39)	Overcure/resin problems	Cutting-free of stresses (see Sec. 3.6.5), tests for overcure (see Sec. 3.8.2), brittleness	Standard tests for overcure and undercure (e.g., AS/NZS 4266.21) not very helpful with unknown materials

TABLE 4.2 Possible Interpretations of Crack Patterns in Melamine Surfaces (HPL, LPL, etc.) and Methods for Further Investigation (*Continued*)

4.3.3 Checks/Cracks in Solid Wood

Checks in wood result from the relief of stresses that are either created by tree growth, for example, as with spiral or interlocked grain, or set up during seasoning (air or kiln drying). The pattern of these "natural" checks often allows them to be easily distinguished from cracks that are due to excessive mechanical loads, for example, during tree felling or machining.

Shrinkage-induced checks virtually always run in the direction of the wood rays (Fig. 4.40). A second natural plane of cracking runs perpendicular to the wood rays at the boundaries of annual rings. The longitudinal separations of the wood found here are termed "shakes" and primarily occur in species with significant differences in density between earlywood and latewood, for example, as with many softwoods from temperate climate zones. Crack development in this direction normally results from exposure of the wood to weathering or mechanical loads (Fig. 4.41). Wavy cracks are occasionally found with interlocked-grain species such as sapele, sipo, bilinga, or iroko (Fig. 4.42). Highly irregular crack directions may also be encountered with some temperate-zone species such as elm. Ultimately, then, the crack direction must always be assessed in relation to the particular wood species or affected component. Any doubts can be resolved by investigating the splitting behavior of the relevant piece of wood. The direction in which it is easiest to split the wood is also the natural direction of cracking.

In contrast, cracks caused by mechanical action, for example, during felling or due to improper working, can run in any direction (Fig. 4.43).

Cracks in solid structural timbers often appear dramatic (Fig. 4.42). However, longitudinal cracks hardly ever impair structural performance as the design calculations already factor in typical wood features such as knots and checks. At the same time, conspicuously oblique cracks may indicate problems connected with the slope of grain (see Sec. 5.2).

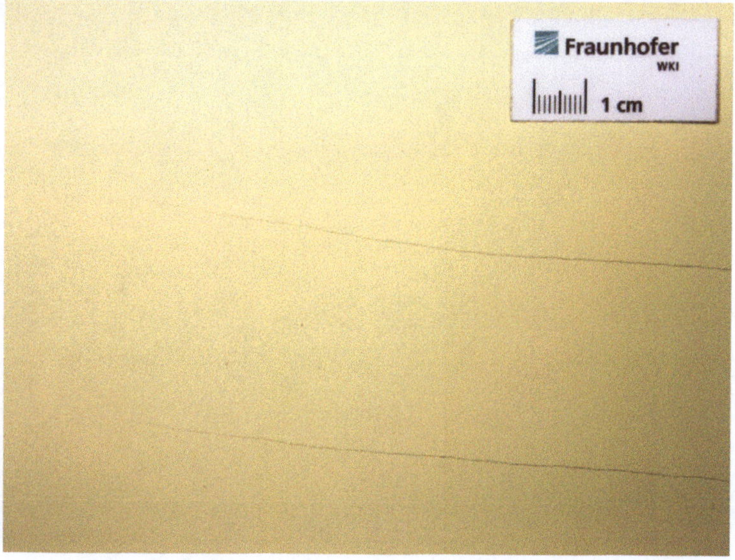

FIGURE 4.36 Parallel cracks in HPL caused by unacceptably shrinkage at lower relative humidity. Due to the high stiffness of the UF adhesive, the cracks formed at intervals of several centimeters. With adhesives of higher flexibility a larger distance between the cracks is to be expected.

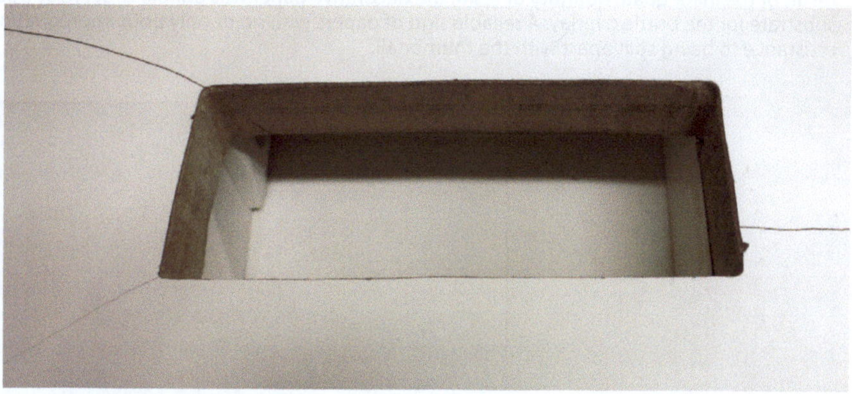

FIGURE 4.37 Although the corners of the cut-out are slightly rounded, two cracks in the CPL originate from the opening. The crack between the right-hand corners, however, confirms that geometrical stress concentration was not an important factor here.

Reaction Wood, Juvenile Wood, Annual Ring Orientation

The inner approximately 5 to 20 annual rings of a tree trunk (juvenile wood) have a lower density and lower resistance to wood-destroying fungi than mature wood. Juvenile wood also displays greater swelling and shrinkage in axial direction than mature wood [54]. Another, special type of woody tissue is formed by trees in response to heavy compressive and tensile loads. This is termed "compression wood" or "tension wood," known

FIGURE 4.38 Irregular cracks in HPL overlay. The area on the left, which had been covered over during service, appears to be free from cracks. A closer look under the magnifying glass, however, also revealed the presence of cracks in the protected surface. As these are not highlighted by dirt, they are hard to recognize with the naked eye. The pattern of the cracks indicates that they are limited to the surface of the HPL. Poor impregnation of the decorative paper layer was identified as the cause. As a result of the inadequate quantity of resin, the decorative paper was unable to serve as a stable substrate for the brittle overlay. A reliable sign of papers with no or only poor impregnation is their low resistance to being split apart with the thumbnail.

FIGURE 4.39 Irregular cracks in a melamine-faced board (MFB) with decorative overlay. Oblique illumination reveals the raised crack edges. Possible explanations for this failure pattern include poor adhesion to the board, splits in the middle of the paper or, as in this case, cracks that extend into the board with turned-up edges. The shape and distribution of the cracks indicate that shrinkage is not the only problem: resin overcure is also thought to have been a contributing factor.

Figure 4.40 Cracks in beech countertops. Pictured on the left is an irregular crack running obliquely to the wood rays, which was probably caused by external action. The right-hand image shows a typical seasoning check running in the direction of the wood rays (radial direction).

Figure 4.41 In typically interlocked-grain species (here bilinga), the crack lines can be heavily undulating (photo by Gerald Koch, Thünen Institute of Wood Research).

collectively as "reaction wood." Compression wood forms in softwoods and is noticeable by its darker color. Tension wood, which is found in hardwoods, is only indirectly observable through eccentric growth, woolly surfaces, and a greater proneness to cracking, but it is easy to detect under the microscope (see Fig. 3.93). Reaction wood, like juvenile wood, exhibits lower strengths and a higher degree of swelling and shrinkage, making it more susceptible to deformation and cracking. Where appropriate, failure analyses on solid wood should take account of these special types of wood tissue.

Although the position of annual rings is frequently irrelevant in present-day wood construction, a number of factors should be borne in mind in the context of failure analyses These include the often superior durability of exterior coatings on the bark side of side-cut boards, the higher incidence of ring shakes on the pith side of flooring products (Fig. 4.43) and, of course, the variations in swelling/shrinkage behavior between vertical- and flat-grained lumber.

FIGURE 4.42 Cracks in white-painted timbers are particularly noticeable. The problem is nonetheless purely cosmetic as the loadbearing capacity is not impaired.

FIGURE 4.43 The mirror-image position of the knots indicates that the two planks originate from the same log. Only the bottom plank, whose pith side faces upwards, shows separations (shakes) between the annual rings. Both planks were subject to the same mechanical loads by chair castors.

Compression Failure

Unusually abrupt fracture patterns in wood components are a typical sign of either fungal decay or compression failures. Compression failures occur where layers of cells are crushed due to excessive longitudinal compression of the wood in the standing tree or overloading during service. The presence of compression lines in wood considerably reduces its strength and, above all, its dynamic load capacity. Pronounced compression failures are often visible on planed surfaces as fine white lines. Less distinct failures, on the other hand, can only be detected by microscopic examination.

Even slight compression failures, visible only under a microscope, may seriously reduce strength and cause brittle fracture. Because of the low strength associated with compression failures, many safety codes require certain structural members, such as ladder rails and scaffold planks, to be entirely free of such failures. [53]

In standing trees, compression failure can, among other things, result from strong wind loads. With spruce, the thin white lines typical of pronounced compression failures are normally spaced at intervals of between 20 and 50 cm [52]. They are chiefly found in the outer zones of the trunk. Very old tropical trunks, on the other hand, often exhibit these lines at the center, only a few centimeters apart (brittle heart).

Compression lines can also form near to wood failures and typically only a few millimeters from the fracture. On the other hand, the spacing of compression lines occurring in installed wood members due to regular compressive overloading is less predictable, that is, is not limited to a few millimeters from a fracture zone or the typical 20 to 50 cm intervals for standing spruces. Hence, knowledge of the compression failure characteristics of the particular wood species under investigation may allow the cause of the compression lines to be narrowed down (Fig. 4.44).

4.3.4 Stress Concentration

Optimum stress distribution is achieved where the stresses present in a member are roughly equal at all points. The shape of trees and the transitions between branches and

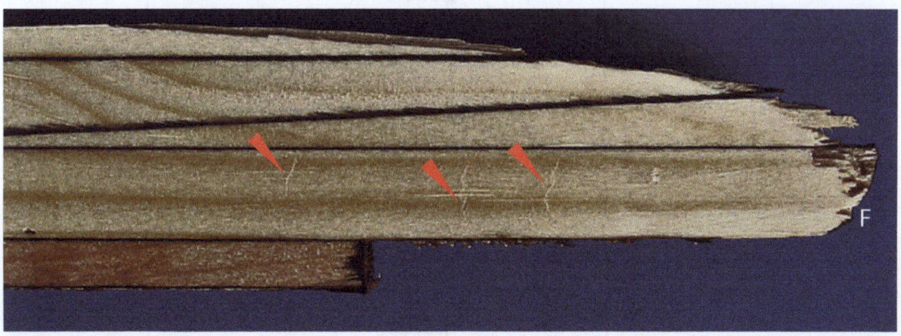

FIGURE 4.44 The smooth-planed surfaces of the fractured wing of a sports plane exhibit compression lines, which are visible with the naked eye on the bottom layer of the support. As their distance from the fracture (F) indicates, the compression lines in the spruce formed neither in the course of the fracture nor in the standing trunk. They are evidently the result of regular overloading in service, for example, during hard landings.

trunk are prime examples of how exceedingly high strengths can be attained by avoiding stress peaks [59]. Critical situations may arise where stresses are concentrated at a single point and no provision is made to accommodate them. Even seemingly harmless grooves or notches can bring about extremely high stress peaks. As the following diagrams illustrate, the notches themselves are the critical factor, rather any resulting reduction in cross-sectional area.

A structural member loaded in tension is subject to the same stress level at every point of its cross-section. Notching, however, leads to irregular stress distributions and a disproportionate loss of strength. Consequently, a 20-mm-diameter circular section, for instance, will offer a much higher tensile strength than a 30-mm-diameter circular section with a 5-mm notch. So-called stress concentration factors have been empirically determined as an aid in evaluating this notch effect in structural members. Notches with a rounded base have by far the lowest impact on strength values while V-shaped notches bring about the greatest strength reduction (Fig. 4.45).

The concentration of stresses (shrinkage and notch stresses) on stress-sensitive components can result in cracking or deformation. Typical cases include cracks originating from rectangular cut-outs, for example, in HPL or glued-laminated timber (Fig. 4.46).

Stress peaks also occur at the base of grooves. The potentially dramatic consequences of geometrically induced stress concentrations at critical locations are illustrated by the example in Sec. 5.9.

A good bond transmits stresses uniformly from one layer to the next. The failure of a bond may cause stress peaks to occur at unexpected locations. Figure 4.47 demonstrates

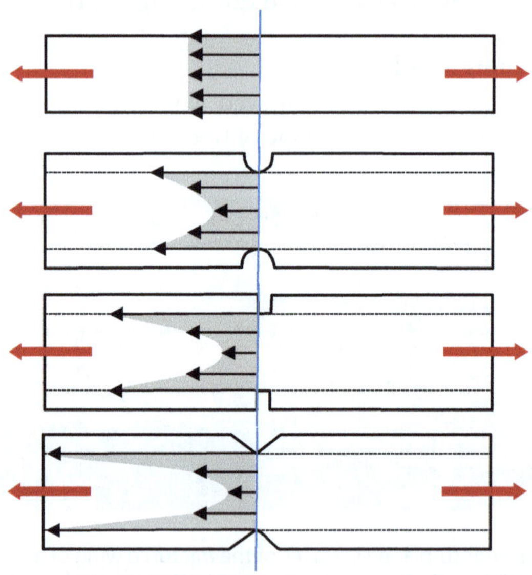

FIGURE 4.45 Impact of notches and notch geometry on stress distribution. Despite their larger cross-section, the bottom three members offer a lower tensile strength due to the stress peaks at the notch location.

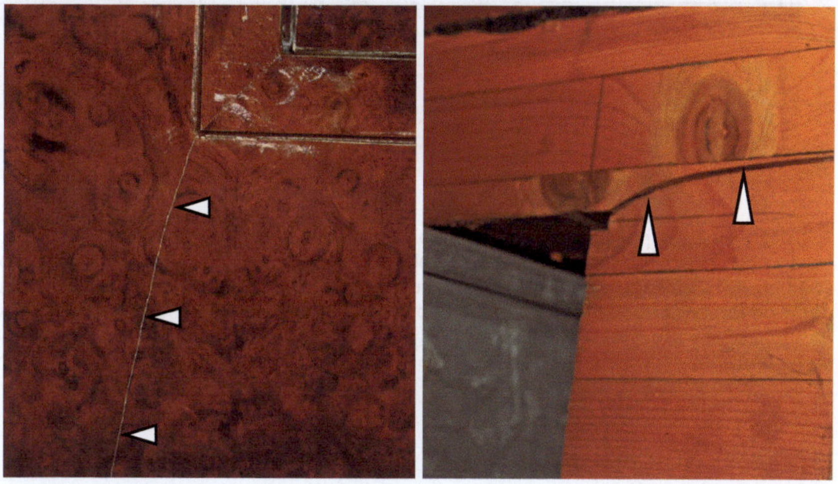

FIGURE 4.46 Left: Crack in HPL originating from a rectangular cut-out without rounded corners, following a prolonged period of extremely low relative humidity. Right: The opening cut in a glued-laminated beam for a ventilation duct penetration was not rounded at the corners. A crack can be seen to originate exactly from the corner. With a neighboring beam, a similar absence of rounded angles at a penetration resulted in full-scale fracture.

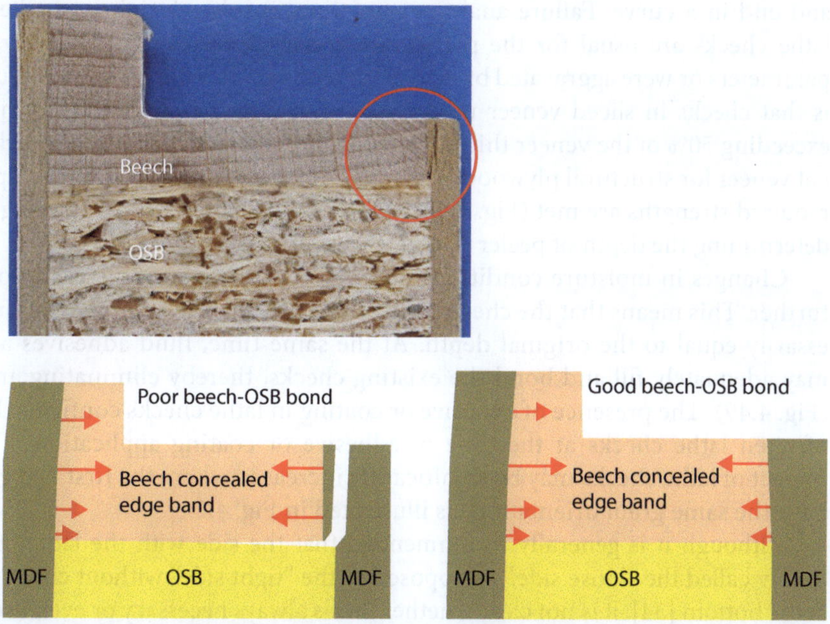

FIGURE 4.47 Cross-section of a door with beech concealed beech edge bands. Cracks in the MDF facings of internal doors were caused by poor adhesion between the shrunken beech concealed edge band and the OSB frame. If the components are properly bonded, the tensile stresses acting on the MDF facing at the junction between edge band and OSB will be negligible. Poor adhesion between edge band and OSB will subject the MDF panel to high perpendicular tensile stresses at the corner of the junction between edge band and frame, with cracking of the panel as the consequence.

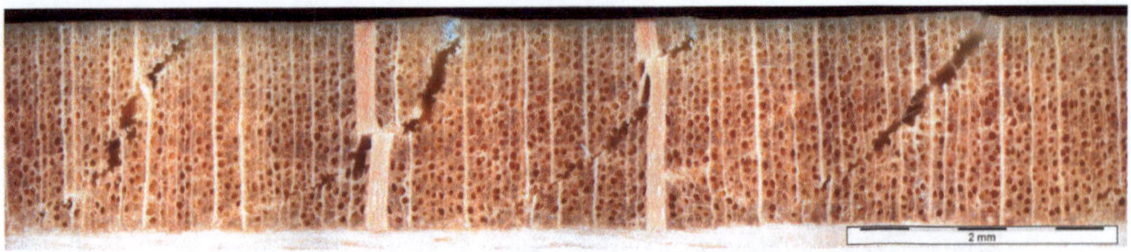

FIGURE 4.48 Deep peeler checks in beech plywood used as a step. The very thin coating, which has hardly penetrated into the checks, was unable to bond the checks adequately and thereby reduce the swelling and shrinkage movement. Given the varying requirements placed on structural plywood and plywood for furniture with regard to the depth of lathe checks, the use of structural plywood for furniture or any other decorative surface may cause problems, regardless of its initial surface appearance.

the consequences of bonding failure for the stress concentrations at the edge of the MDF facing of interior doors with beech concealed edge bands.

4.3.5 Lathe Checks

Lathe checks on one face of veneers are an unavoidable result of the manufacturing process. With rotary-cut veneers, they are often referred to as "peeler checks." Seen in cross-section, lathe checks typically cut into the surface of the veneer at an oblique angle and end in a curve. Failure analyses sometimes involve deciding whether very deep lathe checks are usual for the particular product, were caused by faulty production parameters or were aggravated by adverse service conditions. The general rule of thumb is that checks in sliced veneer used to face furniture items should not have a depth exceeding 50% of the veneer thickness while the occurrence of peeler checks in rotary-cut veneer for structural plywood is not subject to any general restrictions, provided the required strengths are met (Fig. 4.48). AS/NZS 2098.6 describes a staining method for determining the depth of peeler checks in veneer and plywood.

Changes in moisture conditions may cause existing lathe checks to open up still further. This means that the check depth observed during the investigation is not necessarily equal to the original depth. At the same time, fluid adhesives and coatings may adequately fill and bond the existing checks, thereby eliminating any problems (Fig. 4.49). The presence of adhesive or coating in lathe checks confirms the existence of open lathe checks at the time of adhesive or coating application. The negative impact of lathe checks may be significantly increased where the first two veneer layers have the same grain orientation, as illustrated in Fig. 4.2.

Although it is generally recommended that the side with the lathe checks (commonly called the "loose side," as opposed to the "tight side" without checks) be located at the bottom [34], it is not clear whether this is always necessary or even useful. Indeed, tests performed on rotary-cut maple veneer even suggest the opposite [23, 55]. With the loose side bonded, every sanding brings the surface closer to the lathe checks, whereas with the tight side bonded every sanding reduces the depth of the lathe checks. In the case of book-matched face veneers, the loose side and tight side alternate, and both sides have to be fit for purpose.

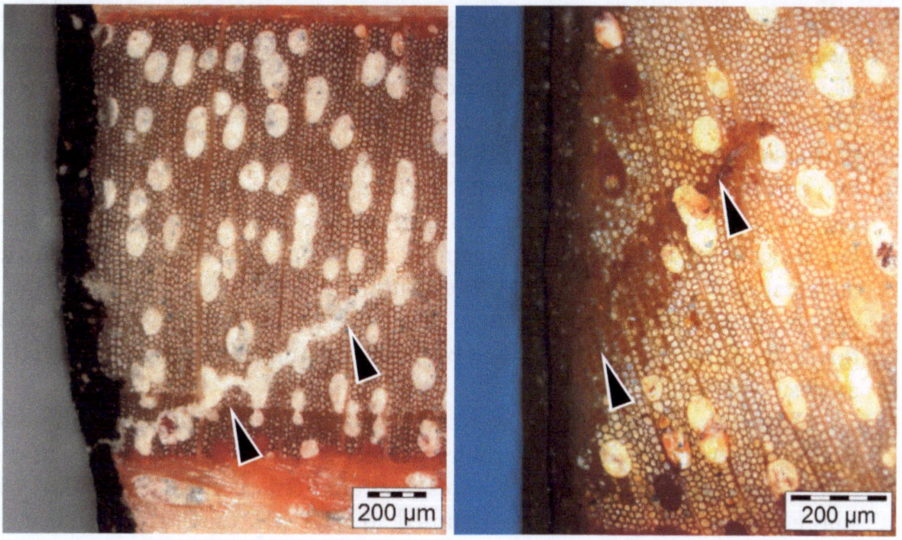

FIGURE 4.49 The crack in the edge of a piece of plywood furniture for outdoor use originates from a lathe check (left). The use of a low-viscosity epoxy resin primer to fill these checks (right) achieved a substantial improvement in coating durability.

4.4 Interpretation of Bonding Failures

Polyvinyl acetate (PVAc), urea formaldehyde (UF), melamine formaldehyde (MF), one- and two-component polyurethanes (PU), epoxies and acrylics are used to both bond and coat wood and other materials. Coatings differ from adhesives due to the absence of pressure during curing, higher aesthetic requirements, the possibility of curing by UV radiation, and exposure to sunlight during service.

Broadly speaking, the proper bonding of adhesives and coatings is subject to the following requirements:

1. The adhesive or coating must be suitable for the application.
2. The material must be undamaged (specific risks include faulty batches, overaging, and frost damage during transportation or storage).
3. The individual components of two-component systems must be correctly proportioned.
4. The substrate must be clean, dry, and strong and contain no substances that inhibit curing or damage the adhesive or coating in the long term.
5. The correct quantity of adhesive or coating material must be applied using an appropriate method.
6. The requirements governing time, temperature, and pressure for drying/curing must be observed.
7. The range of allowable service conditions must not be exceeded.

The substrate for both adhesives and coatings must be clean and undamaged by mechanical or chemical action. Waterborne adhesives and coatings will not, for example, form a durable bond on pitch-covered surfaces such as those encountered with pines, Douglas fir or larch. Oils and other substances from tropical hardwoods may be particularly troublesome, teak being a notable example. The water-repellent pitch/oil prevents wetting and penetration of the adhesive or liquid coating [34]. Traces of migrating pitch, in particular, may start to inactivate the surface of softwood species after the last abrasive treatment (sanding/planing). As this process then starts immediately, it is advantageous to bond or coat surfaces within 24 hours of this treatment. Pitch is often easy to detect due to its distribution near pitch pockets and resin channels. In case of doubt, the situation can be clarified by iodine fuming (see Fig. 3.132) or FTIR chemical analysis (see Sec. 3.8.3).

Chemicals introduced by wood treatment may also interfere with adhesive and coating performance. Familiar problems include the effect on certain reactive PU adhesives [72] and PU coatings of the residual ammonia content in smoked oak. Wood impregnation can generally be assumed to alter pH values and buffering capacity. Impregnated woods may thus pose problems for those adhesives which, for curing, require a certain pH value—regardless of whether this is normally supplied by an acidic/alkaline curing agent or the wood itself. The example in Fig. 4.32 illustrates how the performance of a melamine-urea-formaldehyde (MUF) adhesive is impaired by a fireproofing agent used to impregnate the veneers prior to bonding. Once the failure cause had been identified, it was found that satisfactory bonding could be achieved by significantly shortening the open time. In many cases, trial bonds or coatings with varying process parameters can be investigated as a means of troubleshooting this type of problem. Most importantly, such trials help to establish whether good bond or coating performance is at all achievable. More extensive investigation of such cases is possible through the measurement of pH values and buffering capacities or, possibly, by thermoanalytical evaluation of the reaction speed.

Surface soundness is important for both adhesives and coatings, although most adhesives may be less sensitive to superficial damage as they usually have a deeper penetration and, in the case of UF, MF, MUF, and PF, the ability to stabilize wood cells [33]. Nevertheless, blunt planing knives and excessive pressure during sanding may crush the wood cells to a depth that causes adhesive or coating failure. The first indicator for damaged wood cells is wood fibers that stick to the back of the failed adhesive or coating. With adhesives, however, failure in the wood may be due to a combination of a strong bond and overloading. For this reason, the possible occurrence of damaged wood cells in areas that have not (yet) fallen apart should be examined under the microscope.

4.4.1 Bonding Failure in Joints

Apart from simple detachment, typical symptoms of failing bonds in joints include swelling (Figs. 4.50 and 4.51), offsets, blisters (Fig. 4.52) and cracks (Fig. 4.30).

In many applications, the adhesive not only joins together materials, but also acts as a sealant to prevent the ingress of water. For this reason, a local absence of adhesive may

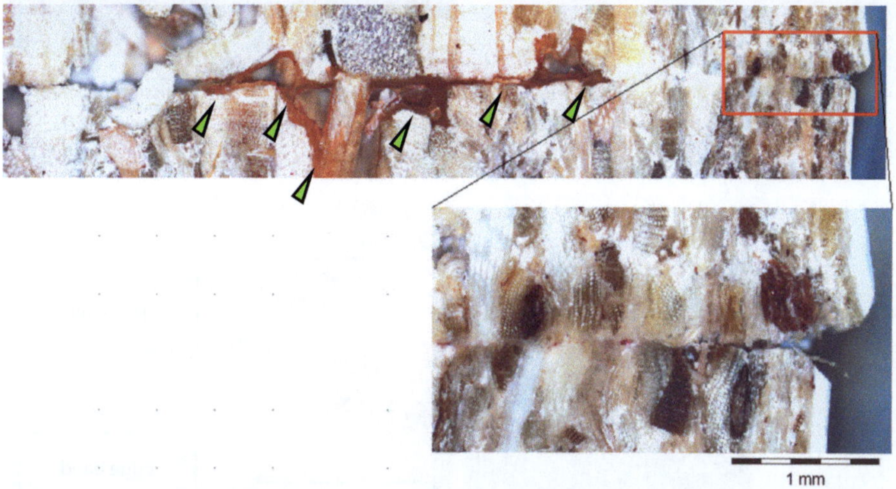

Figure 4.50 Cross-section through joint between two coated particleboard panels in a kitchen. The PVAc adhesive, stained brown with Lugol's iodine, is only present at the panel centers (arrows). Due to the absence of adhesive at the edges, where it would have served an important sealing function, the panels have swollen.

lead to swelling in situations where water or water vapor can penetrate into the assembly. This, in turn, means that poor adhesion can also be signaled by swelling. Typical examples where the sealing action of adhesives is particularly important include kitchen countertops, window sills, and window corner joints. Another frequent function of adhesives is to ensure the uniform transmission of stresses existing between layers.

A question that typically arises in the course of failure analyses relates to whether adhesive bond failure is the result of unsuitable service conditions or poor bond performance. Adequate bond performance in solid wood and wood-based products is normally defined in terms of bond strength, which is required to be at least as high as the strength of the adherends under the projected service conditions. Despite this, the particular circumstances of a failure are still difficult to evaluate sometimes due to the lack of clear-cut requirements for many bonding applications.

The initial assessment of a failed adhesive joint focuses on the location and distribution of the failures together with associated findings such as cracks (or their absence) in the surrounding areas. With veneers or the top layers of parquet flooring, for example, both very wide cracks and cracks with raised edges can be symptoms of poor adhesion (see Sec. 4.3). At the same time, bonding failure in this type of product without any cracking provides an important clue: it suggests that the shear strength of the adhesive is lower than the tensile strength of the wood perpendicular to the grain. With a good bond, however, the tensile strength of the wood perpendicular to the grain is lower than the strength of the adhesive. This explains why, in the case of severe overdrying, for instance, small cracks can be expected to appear in the wood prior to any bonding failure. The absence of small cracks of this kind thus provides an indication of poor adhesion (Fig. 4.53).

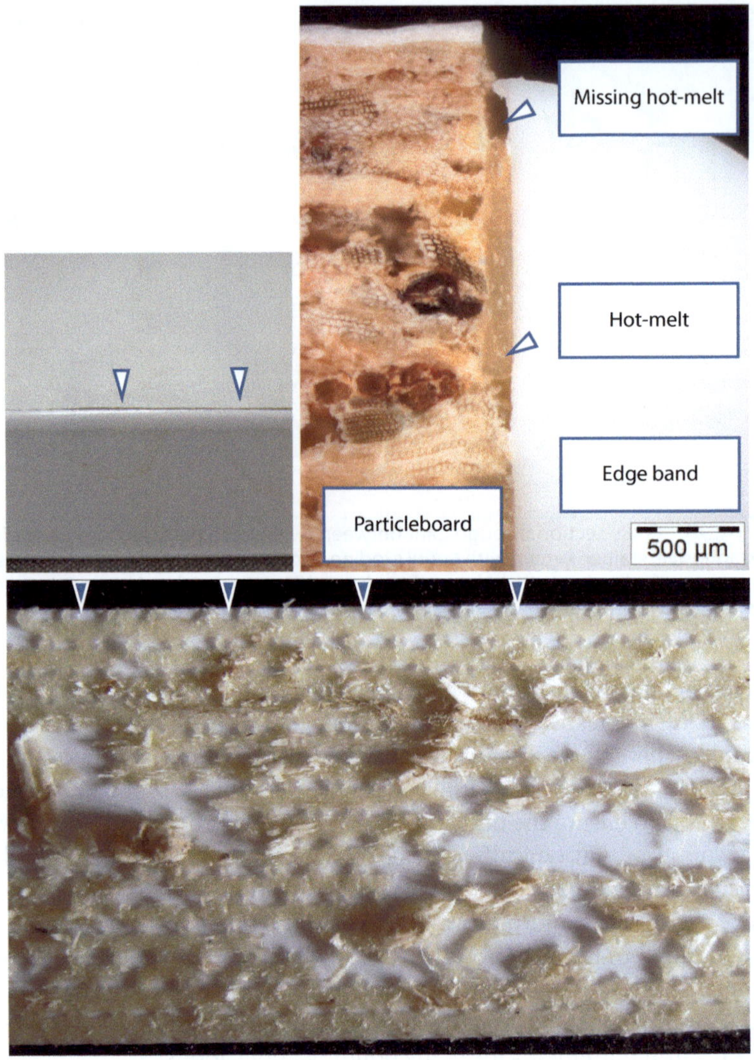

Figure 4.51 Slight swelling is observable in the base panel at the bottom edge of a kitchen front. The cross-section through the swollen edge zone of the particleboard shows an absence of adhesive along the edge band border. This weak point allowed water and water vapor to penetrate into the particleboard substrate, with local swelling as the result. The lack of adhesive is also easily recognizable on the detached edge band.

The explanation of bonding failure that is easiest to verify—and admittedly the most banal—is the local absence of adhesive, a defect that occurs with surprising frequency. Here, the only additional clarification needed is that the service conditions were not responsible for washing out or destroying the adhesive as is sometimes the case, for example, after prolonged moisture action on urea resins.

Excessively thick bondlines, particularly with waterborne adhesives, are liable to cause cracking in veneers and paper-based laminates. Very thin joints, on the other

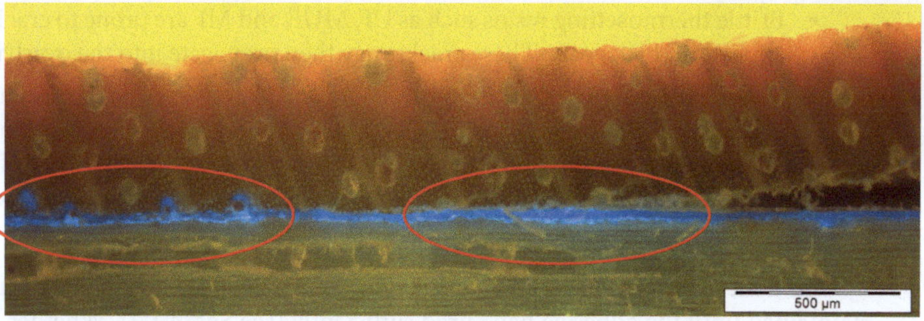

FIGURE 4.52 In the area of good adhesion (left), the adhesive, which glows blue under UV fluorescence excitation, is observed to have penetrated into the adjoining pores of the maple veneer. The edge pores seen in the center and on the right, on the other hand, have not been filled. The bond here is poor and the veneer is peeling away (right). For this reason, it is a fair assumption that the pressure within the bondline was not uniform.

FIGURE 4.53 The complete lack of cracking in the ash top layer of this engineered parquet provides a good indication that the principal cause of failure is more likely to be poor bond performance than overdrying during service [56].

hand, tend to be harmless, provided the adhesive is spread uniformly and adequate pressure is applied up to sufficient curing. Past failure analyses have repeatedly shown that, where joints exhibit a continuous bondline, however thin, the quantity of adhesive should initially be assumed to be adequate.

Thick bondlines may cause cracks in veneers and paper-based laminates through a variety of mechanisms:

- With waterborne adhesives (e.g., UF or PVAc), the veneer or paper may swell strongly before the adhesive has cured. Shrinkage cracks may then develop when the layer dries.

- Brittle thermosetting resins such as UF, MUF, and MF are prone to cracking when applied in thick coats. These cracks may then propagate into the overlying layers.
- Very elastic adhesives, for example, PVAc and isoprenes (e.g., in spray adhesives), are susceptible to creep, especially when applied in thick coats. Dimensional changes in the substrate or bonded laminates can then evidently lead to cracking—also, it would seem, because movement in the bondline is never completely uniform.

The honey-on-toast texture is a typical symptom of failures where the pressure was not maintained until the adhesive had cured sufficiently (see Sec. 4.1.2). Whether or not the pressure was adequate can frequently be decided by the depth to which the adhesive has penetrated into the wood structure. In the case of a properly executed adhesive joint, at least all depressions in the wood surface will be filled with adhesive. With hardwoods, for example, the adhesive can be assumed to fill the open pores immediately bordering the joint, provided these are not blocked by tyloses or heartwood extractives. The full penetration depth of the adhesive into the wood largely depends on the species. This depth is also influenced by the slope of grain of the wood member (see Fig. 5.9). In hardwoods, pores that are blocked by tyloses or heartwood extractives prevent the adhesive from penetrating into adjacent pores. Pores in the heartwood of white oak, for example, are always plugged by tyloses, so only a very low penetration depth is likely for this species. Conversely, with species whose pores are not normally blocked, such as beech or maple, the adhesive can be expected to penetrate deeply into the neighboring pores. An unexpectedly low penetration depth for the relevant species may be a pointer to inadequate pressure, excessive adhesive viscosity or too long a waiting time between adhesive application and pressing (Fig. 4.52).

Uncompressed adhesive beads may be a further indication of insufficient pressure during adhesive curing. Examples of this are presented in Sec. 5.4. Inadequate penetration depth can also be explained by various other factors, such as an excessively long waiting time between adhesive application and the start of pressing.

A good bond is recognizable by flowing transitions between cross-banded layers subject to swelling while an abrupt transition between layers signals poor adhesion (Fig. 4.54).

An important decision in assessing the causes of bonding failure relates to whether the fracture is located in the bondline or the wood (Fig. 4.55). Everyday practice has shown how difficult it often is to decide whether a slight covering of wood fibers on the adhesive should be classed as a failure in the adhesive or a failure in the wood (Zone 4 in Fig. 4.55). It may be useful, in such cases, to apply the distinction drawn between shallow and deep wood failure (ASTM D 5266-99):

Deep wood failure: Failure that is invariably several too many cells away from the adhesive layer, in which the fracture path is strongly influenced by the grain angle and the growth-ring structure.

Shallow wood failure: Failure that is invariably within the first one or two layers of cells beyond the adhesive layer in which the fracture path is not influenced by the wood-grain angle or growth-ring structure.

Table 4.3 lists key causes of failing bonds that have to be confirmed or discarded during a failure analysis together with the techniques that may be applied.

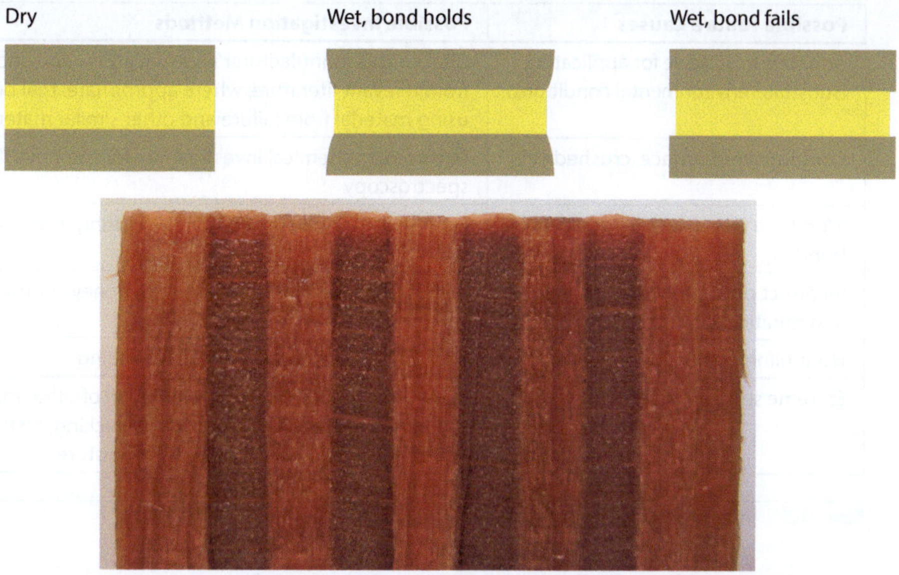

FIGURE 4.54 The state of the transitions between swollen cross-banded plies can help to assess whether the adhesion has failed. Even after 6 hours' boiling, the cross-section of a beech plywood product still shows flowing transitions between the plies. The bond is evidently moisture-resistant and of high quality.

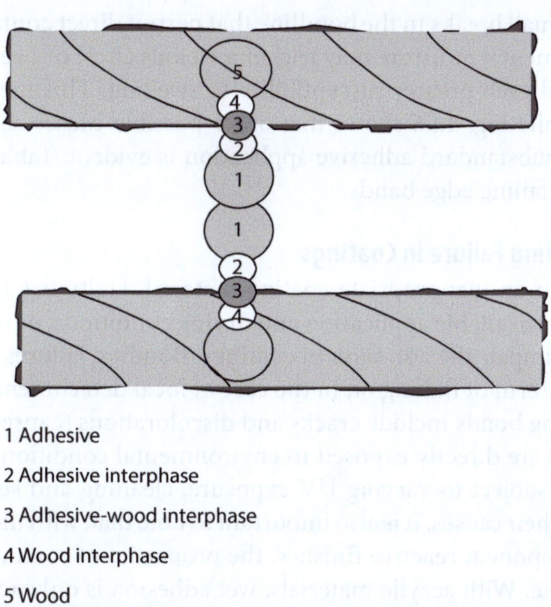

FIGURE 4.55 Imaginary links of adhesive bond between two pieces of wood using schematic from [34].

Possible Failure Causes	Possible Investigation Methods
Adhesive unsuitable for application (substrate/environmental conditions)	Checking of manufacturer's specifications and information from relevant literature; where appropriate, trial bonding using material from failure and other similar materials
Contaminated surface, crushed cells	Microscopy, chemical investigation, for example, IR spectroscopy
Wood too wet or too dry at time of bonding	Estimation of moisture content by cutting free of stresses
Incorrect pressure, temperature, and duration for pressing	Microscopy (wetting, starved joints, honey-on-toast texture, etc.)
Nonuniform spread of adhesive	Microscopy, sanding down, taper sanding
Extreme service conditions	Elimination of other causes, inspection of other materials for damage, inspection of wood for cracking, testing of intact bonds for sensitivity to heat/moisture

TABLE 4.3 Investigation Methods for Failures in Adhesive Joints

4.4.2 Debonding of Edge Bands

Plastic, veneer, and impregnated-paper edge bands are normally bonded with hot-melt adhesives using special "edgebander" machines. The settings for the panel and edge band feed systems, applicator and press rollers, and temperature must be precisely controlled. Cases where debonded edge bands result from the failure to ensure full-surface adhesive application are often relatively easy to diagnose. The achievement of a continuous bondline, at least at the junction with the panel surface, is the minimum requirement here. Small breaks in the bondline that permit direct contact between the substrate and environmental moisture may trigger a vicious circle of swelling, increased substrate exposure and even greater susceptibility to swelling. This mechanism is illustrated in Fig. 4.51, while Fig. 4.56 shows that other possible causes should never be ruled out even where substandard adhesive application is evident. Table 4.4 shows investigation methods for failing edge bands.

4.4.3 Bonding Failure in Coatings

The choice of an inappropriate coating material, faulty batches, inadequate substrate preparation, unsuitable application and drying conditions, or incompatible service conditions may impair the adhesion of coatings. Bonding failures typically manifest themselves in the form of flaking or, in the case of local detachment, blistering. Other initial signs of failing bonds include cracks and discolorations (caused by light scattering).

Coatings are directly exposed to environmental conditions. The pattern of failures across areas subject to varying UV exposure, cleaning and service loads is thus a key indicator of their causes. It is also important to note that, with the exception of UV-curing and two-component reactive finishes, the properties of coatings may change considerably over time. With acrylic materials, wet adhesion is only achieved after several days at temperatures exceeding around 15°C. Particularly under UV exposure, coatings also suffer increasing embrittlement in the course of their service life.

Figure 4.56 Glued side of an edge band that dropped off after a short time in use (top) and the mating plywood edge (bottom). Hot-melt adhesive is absent in a small zone at the corner of the board. Yet, the lack of any signs of swelling or other moisture action suggests that poor application was not the principal cause of the failure. The analysis subsequently showed the extremely low temperature stability of the hot-melt adhesive to be the main factor.

Possible Failure Causes	Possible Investigation Methods
Adhesive unsuitable for application (substrate/environmental conditions)	Determination of adhesive type using FTIR spectroscopy, checking of manufacturer's specifications
Application errors	Visual or microscopic examination: checking for adequate, full-surface adhesive application; checking for penetration of adhesive into voids in panel edge (roller pressure)
Molten adhesive stored for too long/at too high a temperature in glue tank	Inspection of adhesive for brownish or yellowish tinge Chemical/physical investigations
Defective adhesive	Checking of melting point, trial bonding Chemical investigation of interphase (TOF/SIMS) Lack of antioxidant may be signaled by yellowish tinge (consequences of lack of antioxidant for wood-plastic composites (WPC) are described in [50])
Lack of primer on edge band	Fluorescence microscopy (Fig. 4.57), chemical analysis using IR, EDX or TOF/SIMS techniques
Adhesive attacked by resin acids from pine MDF	As yet, problem only encountered with some pine MDF and certain one-component PU adhesives (Fig. 4.58)
Service conditions too wet/too hot	Elimination of other causes, testing of intact edge bands for sensitivity to heat/moisture

Table 4.4 Investigation Methods for Bonding Failures with Edge Bands

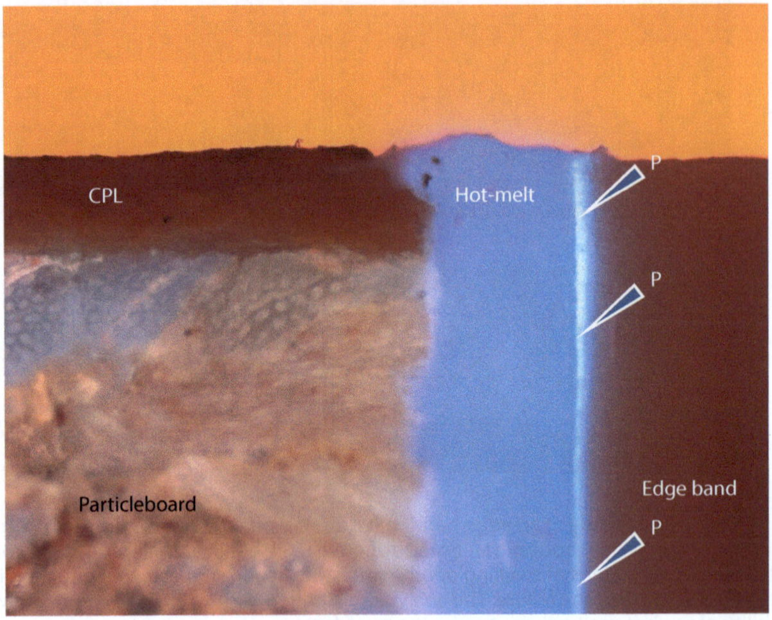

FIGURE 4.57 The presence of a primer (P) is often detectable by UV fluorescence microscopy. If no primer is revealed by microscopic examination, then chemical methods such as FTIR spectroscopy are needed to confirm or to rule out its presence.

FIGURE 4.58 Sometimes, bonding failure of PVC sheets on MDF kitchen fronts occurs only after several years of service. The bondline is evidently susceptible to damage by resin acids from certain MDF grades.

The adhesion and embrittlement (i.e., reduction in the deformability) of coatings is determined by a wide range of factors and it is practically impossible to generalize about specific finish types. Even a standard formulation for white exterior paint contains—apart from the actual binder (e.g., acrylic)—some 8 to 20 additional constituents, whose precise interaction is only known (or, at least, should be known) by the manufacturer.

The problems associated with the measurement of coating adhesion and the assessment of results are discussed in Sec. 3.6.2. The tests above all permit comparative statements to be made about the relative adhesive strength at different locations.

Indoor Applications

Coating adhesion rarely poses problems in dry indoor environments. Failures that do occur often take the form of intercoat peeling, involving a loss of adhesion between the priming and finishing coats. This may result from a failure to perform intermediate sanding, which would have removed troublesome substances, such as a floating wax, from the surface of the priming coat (Fig. 4.59).

Bonding failures in coatings can also be triggered by chemical action, for example, plasticizer migration from adhesive tape that remains stuck on a surface for too long (Fig. 4.60), unsuitable cleaning agents, and reactions involving fireproofing agents or the ammonia from smoked oak.

Interior coatings may also fail because the effective moisture exposure, for example, of baseboards in public buildings and some bathroom furniture, is virtually equivalent to that of outdoor coatings.

Wet Adhesion

Bonding failure is one of the most common problems in outdoor areas, where it is often very difficult to assess. The adhesive strength between coating and substrate tends to weaken considerably upon the ingress of moisture, as the wood swells and water molecules infiltrate below the coating [58]. Detachment of the coating from the wood takes the form of large-area peeling or local blistering. In favorable cases, the adhesive strength may be spontaneously regained after drying. Yet, if the wet adhesion of the coating is too

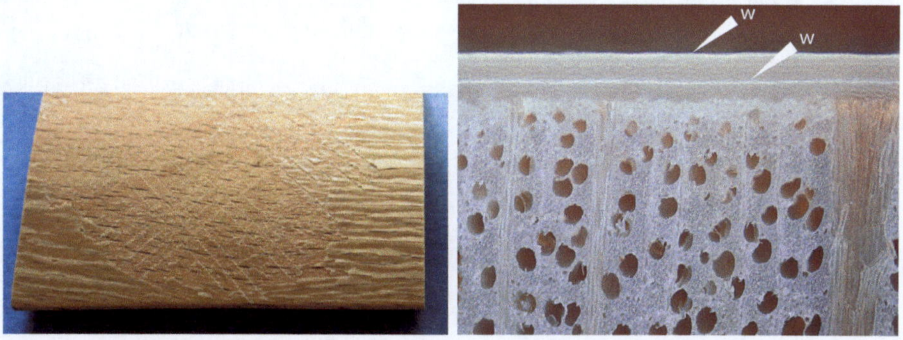

FIGURE 4.59 Intercoat peeling on beech. The failure may be attributed to a thin wax layer (W), visible under a fluorescence microscope, that has floated up to the surface of the priming coat.

FIGURE 4.60 Typical peeling failure caused by adhesive tape that is either incompatible or remains stuck on the surface for too long.

low, even minimal damage to the film can lead to extensive failures. Water penetrating into the wood through such defects can, after only a short time, initiate large-scale bonding failures in the coating.

Coatings with good wet adhesion performance can sometimes be a mixed blessing. With wood windows, for example, excellent wet adhesion can protect against damage during the construction period. At the same time, however, it may help to conceal dangerous levels of moisture penetration. Prolonged moisture penetration at windows, for example, is inacceptable due to its potential impairment of the frame insulation and the risk of decay by fungal attack. From this perspective, it would actually be useful for long-term moisture exposure to become outwardly visible through blistering or flaking of the coating film (Fig. 4.61).

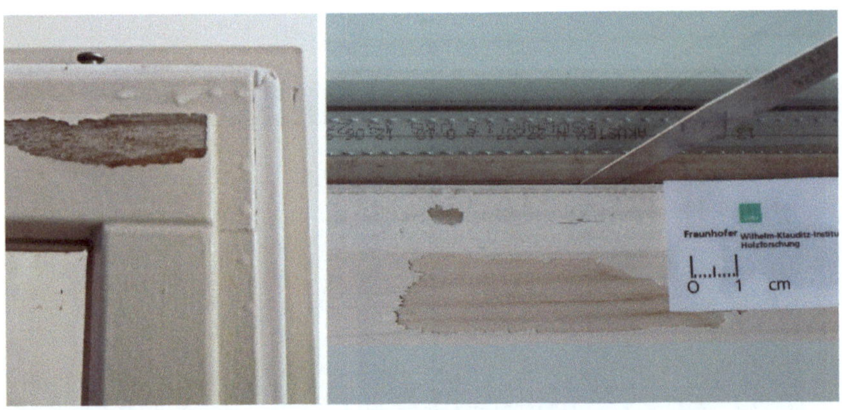

FIGURE 4.61 Loss of wet adhesion after long-term moisture action through condensation caused by inadequate airtightness (left) or moisture ingress due to a failed sealant (right).

In many cases, the wet adhesion of acrylic dispersions is improved by preservative impregnations (e.g., to protect against blue-stain fungi). Wet adhesion can also be improved by additives (wet adhesion promoters) incorporated in the coating formulation by the manufacturer. According to [26], mechanical anchoring plays a key role in the achievement of good wet adhesion for wood coatings. With acrylic dispersions, full wet adhesion is only obtained after several days, even at room temperature. Where temperatures are below around 10°C for several months (e.g., in winter), wet adhesion performance may remain at the original poor level throughout this period. If the coating is then reexamined after a series of warm days, a significant improvement in wet adhesion may suddenly be observed.

Frequently encountered, though fortunately easy to detect, is adhesion loss due to the insufficient removal of previous coatings or damaged wood surfaces during renovation. In this case dirt or mildew/mold, previous coatings or wood fibers from previous UV-induced lignin degradation will adhere to the back of the detached coating.

Some Remarks about Blisters

Blisters are a typical symptom of bonding failure in exterior coatings. Given that high deformability is a precondition for the formation of blisters, their presence clearly indicates that the coating cannot have been very brittle at the time of their appearance. As many finishes embrittle over time, blisters are normally observed on relatively fresh coatings that still display adequate deformability. Older, embrittled coatings tend to fail by flaking rather than blistering.

Further prerequisites for blistering include a local loss of adhesion and a buildup of liquid or gas pressure. For this pressure to accumulate, the finishing coat must exhibit relatively low permeability to the relevant liquid (usually water) or gas (air, water vapor or solvent). Under practical conditions, the pressures that can develop through the heating up of water or air are inadequate to produce blisters in well-adhering coatings. Only with solvent-borne coating systems can gas pressure be considered as a potential cause of blistering. This is because evaporating solvents are able to generate pressures much higher than water vapor or air at temperatures far below 100°C.

As blistering always entails a loss of adhesion to the substrate, all mechanisms that reduce adhesive strength constitute potential failure causes, for example, water infiltration (wet adhesion) (Fig. 4.62), adhesion-reducing substances such as unsuitable surface fillers (Fig. 4.63) and pitch (Fig. 4.64), or mechanical defects, for example, insect burrows (Fig. 4.65), cracks/checks in the wood, and holes, cracks or imperfections in the priming coat.

Blisters can also form without the action of moisture or solvents where the melting point of thermosoftening primers/fillers is exceeded due to extreme heat exposure, for example, on account of southerly orientation or dark coatings (Fig. 4.66). This failure mechanism is identifiable by the characteristic honey-on-toast texture, see Sec. 4.1.2).

Flaking of Coatings due to UV Damage

Roughly 60% of wood is made up of cellulose and hemicelluloses, both of which are highly resistant to UV radiation. In contrast, lignin, which accounts for the

FIGURE 4.62 Typical blisters in a window coating. The blisters appeared shortly after installation in the cold season, when the fresh acrylic finish had not achieved its full wet adhesion and high levels of moisture prevailed due to the recent performance of plastering and cementitious topping works.

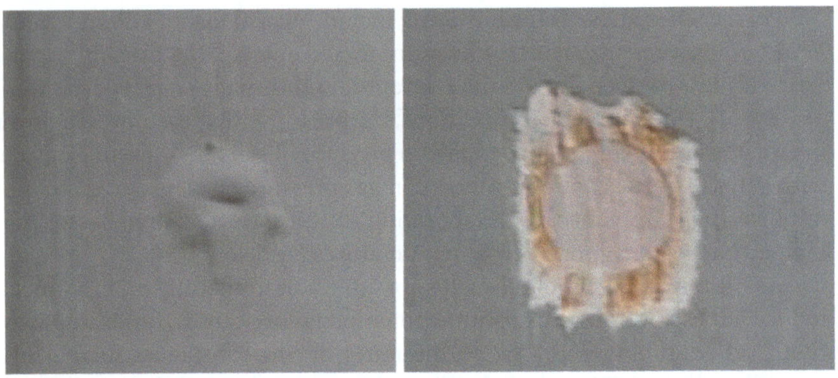

FIGURE 4.63 The blister formed above a drill hole plugged with filler material. Adhesion to surface fillers is often poorer than to wood.

remaining 40% or so of wood and acts as the "cement" between wood cells, can be degraded to a depth of approximately 0.3 mm into small water-soluble fragments by the high-energy, UV component of sunlight. The absence of lignin can be detected by chemical methods and, using suitable staining methods, can also be observed under the microscope. Degraded lignin is also clearly visible with the naked eye or under a magnifying glass on the underside of peeled coatings. Where a covering of wood fibers adheres to the detached film, the wood is most likely to have been degraded by

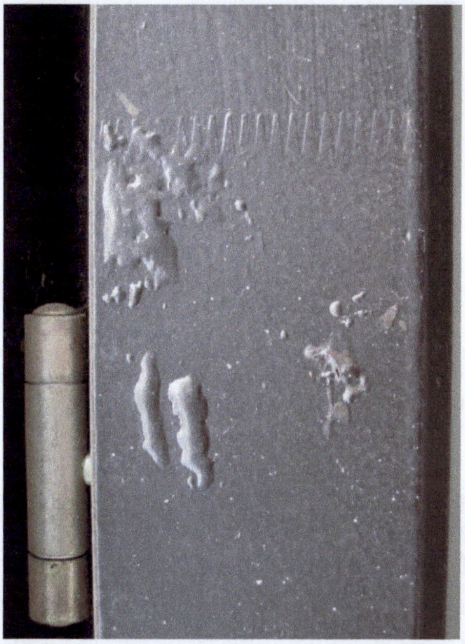

Figure 4.64 Large quantities of pitch brought about a local reduction in adhesive strength sufficient to cause blistering. Even without breaking open the blisters for inspection, it can be safely assumed, due to the absence of failures in the mating component of the finger joint, that local wood features were responsible.

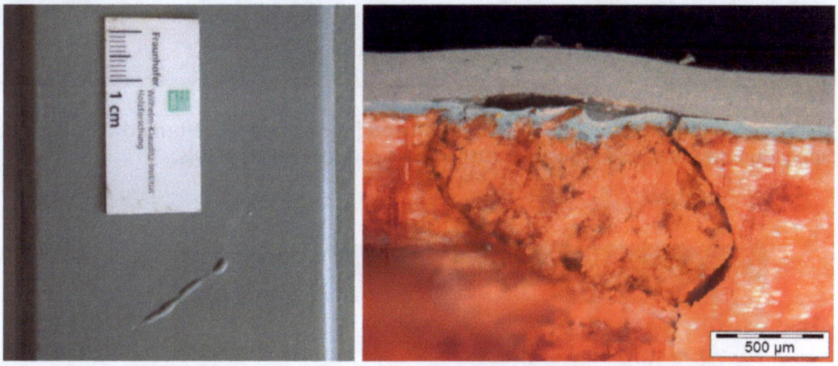

Figure 4.65 Blisters running across the grain can only form above defects in the substrate, patches of filler or, as pictured here, the loose area of an insect burrow. The micrograph of the cross-section shows the burrow clogged with bore dust and the bonding failure in the overlying coating.

UV radiation—provided the possibility of damage by chemicals, the crystallization pressure of salts [49], or extreme crushing of the wood surface (Fig. 4.67) can be ruled out (Fig. 4.68).

Table 4.5 lists key causes of coating problems that have to be confirmed or discarded during a failure analysis together with the techniques that may be applied.

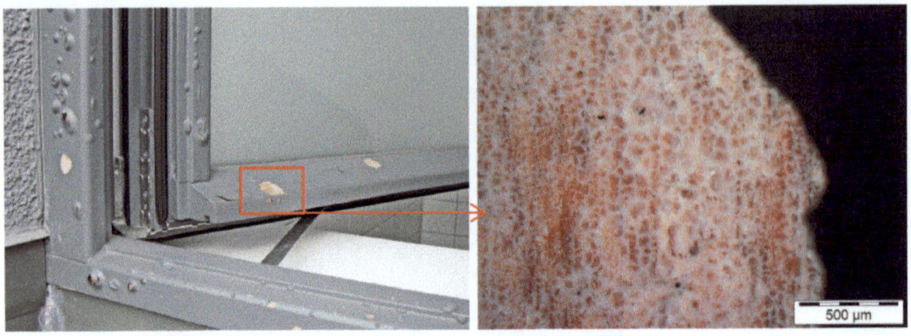

FIGURE 4.66 Numerous blisters formed on a window sash frame. The base of the open blister exhibits the typical honey-on-toast texture. The blisters were limited to surfaces with direct exposure to sunlight and evidently resulted from the oversensitivity of the priming coat to high temperatures.

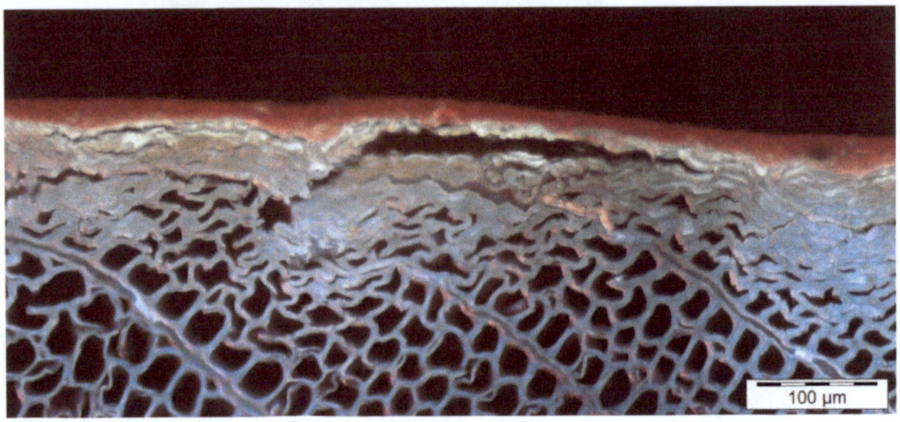

FIGURE 4.67 The cross-section of coated softwood shows heavily deformed cells. It is not possible to apply durable coatings to severely crushed wood surfaces such as this.

FIGURE 4.68 Flaking transparent coating on lower frame member of a window sash. Wood fibers on the rear face of the flakes are clearly visible with the naked eye or under a magnifying glass. This wood fiber covering proves that the coating offered insufficient UV resistance.

Possible Failure Causes	Possible Investigation Methods
Coating unsuitable for application (substrate/environmental conditions) Incompatible coats	Checking of manufacturer's specifications and information from relevant literature; where appropriate, trial coating using material from failure and other similar materials
Crushed surface	Microscopy Checking for adhesion of wood fibers to underside of detached coating
Surface contaminated, for example, with oils or resin, inadequate surface wetting	Microscopy: staining for oils with Sudan IV. Checking for imprint of wood on underside of coating Chemical analysis: IR spectroscopy, GC/MS, TOF/SIMS
Film buildup defective, unsuitable film thickness, separating layers (wax)	Microscopy/fluorescence microscopy Chemical analysis: IR spectroscopy
Inadequate UV resistance	Assessment of failure distribution, checking for adhesion of wood fibers to underside, microscopy with staining methods for lignin (phloroglucinol) Measurement of UV transmission of free films
Faulty coating batch or damage to coating, for example, by frost	Checking for fine cracks at locations with low weather exposure (embrittlement), checking of dry adhesion and wet adhesion on nonexposed surfaces (e.g., on interior faces of windows)
Chemical action	Trial application of same coating on sanded-down substrate
Extreme service conditions	Assessment of failure distribution, checking for damage to other materials, checking of sensitivity of intact coatings to heat/moisture, elimination of other causes

TABLE 4.5 Investigation Methods for Flaking, Peeling, or Blistering of Coatings

4.5 Interpretation of Discolorations

Wood and wood-based products may suffer unwanted color changes at practically any point in the process chain. Various explanations for discoloration include light scattering, UV radiation, chemical or biological action, dyes, extractives, and dirtying. Discolored patches in wood stains caused by slope of grain and bacterial attack are described in Sec. 5.2.

In some cases, the color changes themselves may be the main issue while, in others, they are merely a symptom of deeper-rooted problems. Discoloration is sometimes easily removable just by cleaning. Should it recur, however, it may well be due to some fundamental defect. In many cases, discoloration only becomes noticeable where it creates an unusual pattern of color variation. This is what makes color deviations that run across the grain particularly conspicuous. Complete, uniform changes in color, on the other hand, often go unnoticed.

It is possible, where necessary, to measure the color values of discolored areas using special equipment (Fig. 4.69).

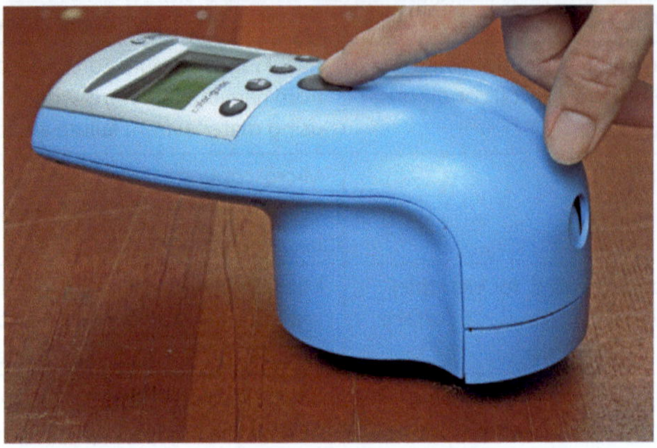

FIGURE 4.69 Exact color values can be determined using color measurement devices.

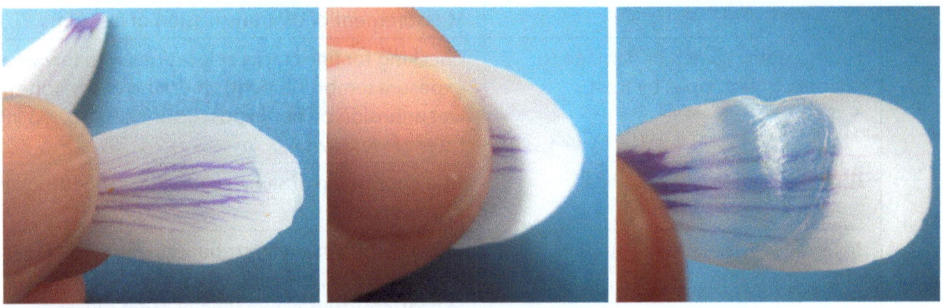

FIGURE 4.70 A crocus petal loses its white color and becomes transparent after being pressed between the fingers. The densification of the cells eliminates the light scattering.

4.5.1 Light Scattering

White sunlight or artificial light passes through transparent coatings and when it hits the substrate, for example, wood, certain wavelengths are absorbed. The remaining light is reflected by the substrate and, from here, reaches the observer's eye. The absorption of certain wavelengths in the original white light and the reflection of others is what lends the wood its color. At boundary surfaces between materials with different refractive indices, for example, very fine cracks, the incident light is scattered and diffusely reflected in its original color (i.e., normally white). The white color is thus produced by imperfections within a layer. A white petal, for instance, is actually transparent and only appears white because the light is scattered and diffusely reflected at numerous boundary layers (Fig. 4.70).

Bonding failures, the edges of cracks and entrapped air bubbles in transparent layers often appear a diffuse whitish or grayish color. Typical examples for this type of light scattering include the discoloration of peeling or blistering finishing coats (Fig. 4.71), the silvery-white appearance of earlywood pores over adhesive bleed-through (Fig. 4.72), stress whitening in coatings (see Sec. 3.6.2) and whitish-colored patches in melamine finishes resulting from local imperfections in the panel, press or coating (see Sec. 5.3, Fig. 5.17).

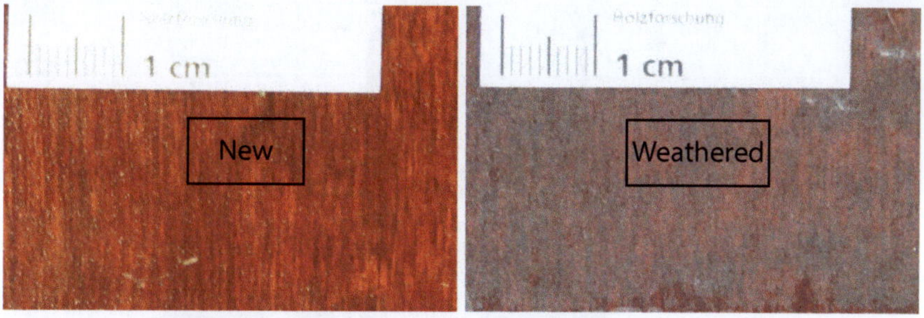

FIGURE 4.71 A laminate surface turned milky gray after a short period of weathering (left: unweathered; right: weathered). The discoloration signaled the start of failing bonding of the overlay.

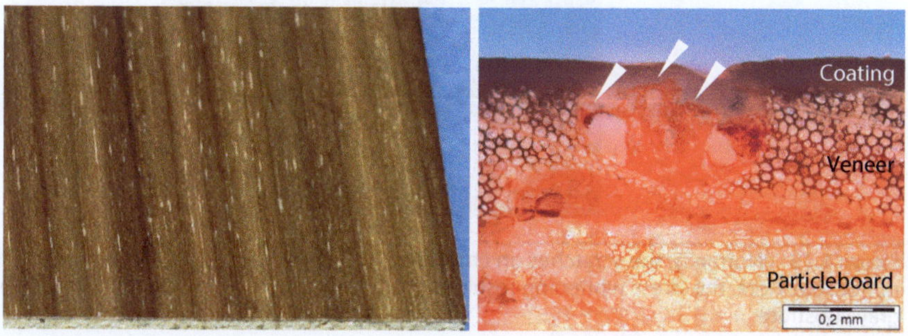

FIGURE 4.72 Whitening in the grain caused by adhesive bleeding. As the micrograph of the cross-section shows, the coating material near the pore in the walnut veneer is in direct contact with the PVAc adhesive, which is stained brown with Lugol's iodine (arrows).

Very fluid adhesives, high pressures, and thin veneers can induce the bleed-through of adhesive onto the surface of veneers. This type of bleed-through primarily occurs above earlywood pores or other structurally loose areas. Discoloration is a typical feature of such failures and is due to the reduced absorption of wood stains, pH value shifts and whitening in the grain (Figs. 4.72 and 5.20). Any whitening in the grain resulting from an incompatibility between adhesive and coating becomes immediately apparent. With brittle adhesives, on the other hand, longer periods may elapse before fine stress cracks show up white through the transparent film.

When subjected to mechanical loads, very brittle coatings undergo a phenomenon called "stress whitening" (Fig. 4.73). This failure typically occurs when brittle coats are exposed to high stresses. Ductile polymer films are not prone to stress whitening as they merely deform without developing new boundary surfaces. With brittle polymer films, on the other hand, stress whitening results from new boundary surfaces created by the fractures within the coating (or between coats).

A white or gray discoloration, known as "blushing," may occur when waterborne transparent coatings cure below the minimum film-forming temperature. A temporary or permanent milky appearance on the surface of coatings may also be caused by overly rapid solvent evaporation or the presence of excessive moisture during the curing process [21].

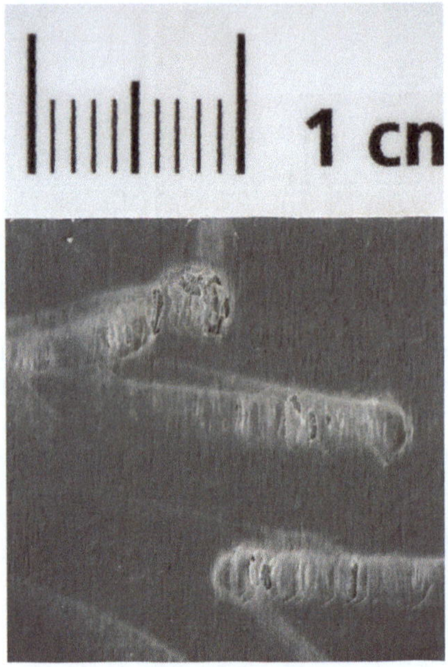

FIGURE 4.73 Normally, the scratches on chair seats in schools caused by studs on trousers are hardly noticeable. Here, the stress whitening resulting from the scratches provides a striking contrast to the gray substrate.

4.5.2 Fading due to UV Radiation

Fading is often the result of exposure to UV radiation. This is because the high-energy UV component of sunlight has a tendency to destroy the natural dyes in dark wood species and the pigments/dyes in stains and other finishes. Particularly unsightly is the selective fading of individual pigment types, for example, a gray veneer which turns green due to fading of the red pigments, without the simultaneous degradation of the blue and yellow pigments.

When exposed to daylight, untreated wood undergoes natural—and sometimes considerable—changes in color caused by both UV and visible-wavelength light [85] (Fig. 4.74). Light-colored woods usually darken while dark woods turn lighter. With many species, this patina effect is desirable as it enhances the inherent charm of the material. To prevent natural discoloration by UV light, wood is either bleached and then treated with a UV-resistant stain or directly finished with a product with built-in UV resistance. The performance of UV-protective coatings is sometimes difficult to assess and any such judgments can only ever apply to specific cases. Moreover, the achieved protection not only depends on the quantifiable UV transmission, but also, among other things, on the quantity of radical scavengers in the finish. The UV transmission of coating products, either in liquid form or as free-film specimens, is simple to determine using suitable spectrometers. Testing for the UV transmission of applied coatings, on the other hand, is more complex and not always feasible given the difficulty of first separating an intact film specimen from the substrate. Specimens are created by bonding the coating to a UV-transparent base with a suitable adhesive and then carefully removing

FIGURE 4.74 The impact of UV radiation is most evident where the color of areas protected from sunlight differs significantly from that of the exposed surfaces.

the wood using sandpaper and a scalpel. The UV absorption of the coating remaining on the base can then be established.

In case of doubt, the UV resistance of coatings can also be determined by measuring the discoloration of unaffected areas after artificial irradiation and, where appropriate, comparing the results with those for other standard commercial products.

4.5.3 Chemical Discoloration

Chemical substances that typically induce color changes include acids, alkalis and oxidizing agents such as hydrogen peroxide. Iron too can be a source of discoloration, due to its reaction with the natural tannins in wood (described in Sec. 5.5), as can the extractives of some tropical hardwoods.

[72] describes how the rubber feet of household appliances or oxidizing agents such as hydrogen peroxide and chlorine can bring about the orange discoloration of aromatic polyurethane finishes.

According to [67], cherry and maple veneers may turn red upon exposure to alkaline adhesives. The discoloration of makore, beech, birch and gaboon by alkaline substances is described in [74]. Illy et al. [41] explains how the aluminum chloride additives used in PVAc adhesives to increase moisture resistance can produce color changes in beech and, above all, cherry veneers. Here, the pressure, light exposure and applied finish were also identified as contributing factors

There are three basic methods for detecting discoloration by chemicals:

1. Direct detection of reaction-initiating chemicals or their reaction or degradation products
2. Indirect detection by reproducing discoloration
3. Interpretation of clues provided by distribution

The detection of chemicals is, in some cases, relatively straightforward: potassium thiocyanate can be used to test for the presence of iron, for example, while the residue of acids or alkalis can be identified on the basis of its pH value (Fig. 4.75). EDX spectroscopy offers an alternative where the use of pH indicators is unsuccessful. Provided the chemical residue has not been removed by cleaning, certain peaks in the EDX spectrum may supply indirect evidence, for example, of hydrochloric acid (chlorine peak) or caustic soda (sodium peak). Volatile acids and alkalis such as ammonia and acetic acid, on the other hand, are no longer directly detectable after a certain period. Here, the investigator is forced to rely on indirect methods.

It is often feasible to reproduce the investigated discoloration, especially where its cause can be narrowed down to only a few substances, for example, cleaning agents. Where the analysis focuses on cleaning agents, for instance, the product can be applied to unaffected surfaces both in properly diluted and in concentrated form (see Sec. 5.6). The example in Fig. 4.76 shows the reproduction of pink discolorations caused by phenol-resin-coated plywood.

Should neither chemical detection nor reproduction appear feasible or worthwhile, it may still be possible to gather clues from the failure distribution (Fig. 4.77).

Apart from other problems, such as the inhibition of curing, the extractives of tropical hardwoods in particular may cause haze and discoloration in transparent coatings (Table 4.6).

4.5.4 Discoloration of Solid Wood and Veneer

To clarify the causes and, possibly, the acceptability of wood discoloration, as well as liability issues, it is important to distinguish between [51]:

- Discolorations that originate from the living tree. A further distinction to be drawn here is between discolorations caused by fungi, bacteria or wound reactions, and thus atypical of sound wood, and discolorations resulting from the localized internal deposition of extractives, which should be regarded as a

Figure 4.75 Brown patches appeared on the slat of a sauna bench. One of the patches was dampened and tested with pH indicator paper. The recorded pH value of 1 indicated that a strong acid was responsible for the discoloration.

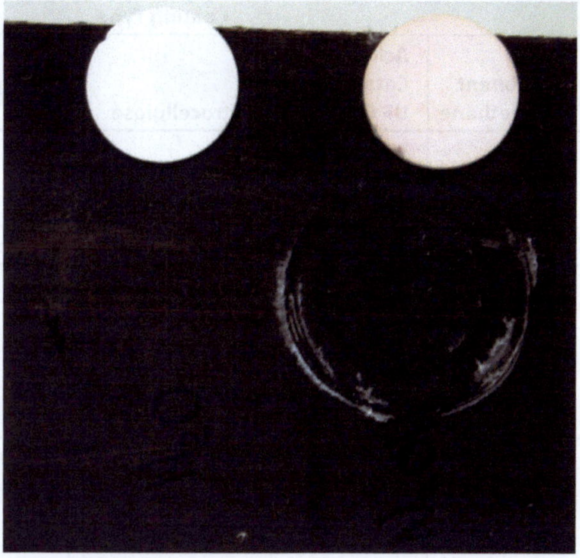

FIGURE 4.76 Phenol-resin-coated plywood had caused pink discolorations on white-colored exposed concrete. A filter soaked in water and placed on the phenol resin film showed no signs of staining. A filter soaked in a caustic potash solution, on the other hand, turned pink after only a few hours. Certain low-molecular-weight phenols are liable to cause this kind of discoloration in highly alkaline environments (such as that created by fresh concrete). The fact that the discoloration ceased to occur after heat treatment of the plywood suggested inadequate curing of the paper laminate as the cause of the problem.

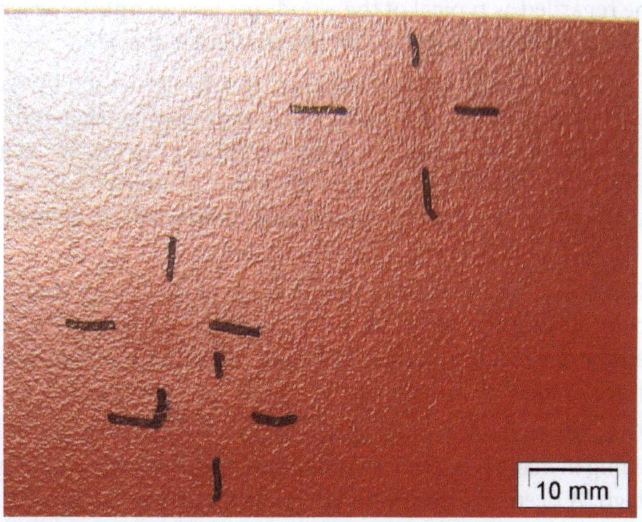

FIGURE 4.77 Vaguely discernible discolorations on high-pressure laminate, which cannot be removed by cleaning. Chemical detection methods cannot normally be used to investigate discolored patches such as this. However, the position of the marks shows a close correlation to the fingerprints of a hand. The surface was possibly touched by acid-contaminated fingers or gloves.

Wood Species	Coating Type				
	Two-Component Polyurethane	Acid-Catalyzed UF Coating	Nitrocellulose	Drying Oil/ Alkyd	Unsaturated Polyester
Spruce, pine, resinous parts of larch					Drying delay
Oak				Discoloration	
Wenge		Discoloration	Exudation		Drying delay
Mansonia				10-day drying delay	Drying delay
Iroko				5-day drying delay	8-day drying delay
East Indian rosewood and other rosewood species				10-day drying delay	73-day drying delay
Teak	Haze	Haze	Haze	Drying delay	5-day drying delay

TABLE 4.6 Effects of Extractives on Coatings: Examples from a Study [75] in Which 91 Species Were Tested

natural phenomenon for the wood species. A judgment on whether a particular form of discoloration caused by particular stress factors during growth should be regarded as typical of the wood species can only be reached on a case-by-case basis, with due allowance for the particular species.

- Discolorations in freshly felled or stored logs caused by reactions involving the still-living parenchyma cells, which may also be triggered by the temperature at the time of seasoning [81]. Given that the presence of oxygen is often a precondition for the discolorations, they are primarily encountered at the ends of trunks. The problem can be solved by killing off the living cells with sulfur dioxide or methyl bromide prior to seasoning [81].
- Discoloration due to improper storage (discoloring fungi/bacteria). Fungal discoloration (blue stain, mold or incipient decay) is the easiest type to detect, either by general appearance and distribution (heartwood/sapwood) or by microscopic examination.
- Discoloration due to improper drying or steaming. "Brown stain in pines and darkening in many hardwoods is a common problem when drying temperatures are too high. A deep grayish-brown chemical discoloration can occur in many hardwood species if initial drying is too slow or too high of an initial kiln temperature" [17].
- Discoloration due to (improper) use, for example, reaction of natural tannins with iron (see also Sec. 5.5) or reaction of extractives with adhesives [41].

The position of the discolorations frequently provides clues as to the cause. Marks that are 10 to 40 mm wide and run across the grain are most likely to be sticker stains, resulting from the contact between the sticker and the board. Drying takes longer below the stickers and there may be less contact with oxygen. Stains that occur predominantly at the ends of boards are a typical sign of oxidative discoloration during storage. With square-sawn timber, the presence of stained patches on all faces indicates that these cannot have existed in the original log prior to conversion. Moreover, where the characteristics of the specific wood species are known, many discolorations can be diagnosed simply on the basis of their color and features. Yet, often, the causes of wood discoloration are complex and, despite decades of intensive research, many questions are still unresolved.

4.5.5 Dirt and Staining Agents

Discoloration by dirt or staining agents requires the continuous presence of agents such as cigarette or candle smoke, exotic spices, red wine or oil, together with the necessary surface properties to absorb or hold the staining substances. Such surface properties include geometrical features (e.g., pores; see Fig. 2.2), surface roughness, surface adhesion or a means of penetration into the surface.

The staining agent may be present in extremely low quantities, making chemical analysis difficult or impossible. As a result, the size and distribution of the discoloration (proximity to hardware, wipe marks/scuffing, presence on adjacent surfaces/materials) are key factors to consider. A second way of identifying the source of the problem is to attempt to reproduce the discoloration with the candidate substances. In industrial environments, all liquids, dusts and surfaces that are present at some stage of production can be collected and tested.

4.5.6 Extractive Bleeding

Color heartwood extractives sometimes migrate through coatings and stain surfaces (Fig. 4.78). If this type of discoloration is uniformly distributed or, like many other wood features, runs along the grain, it frequently goes unnoticed. In the case of finger-jointed members with heartwood and sapwood components, on the other hand, the discolorations tend to be particularly striking as the color change runs across the grain and emphasizes the joint (Fig. 4.79).

With most species, stain-blocking primers can provide fairly reliable protection against extractive bleeding. Finger-jointed pine, however, is a common cause of disputes in this connection (Fig. 4.79). Pine does not usually suffer major problems from discoloring extractives, and abnormal environmental factors, such as high temperatures, moisture or a significant change in conditions, are usually needed to trigger the migration of extractives to the surface. For this reason, the stain-blocking function of the coating is sometimes given too little consideration.

One problem not yet completely resolved concerns knots in pine and spruce that show through as yellowish/brownish spots on white finishes. The discoloration over the knots can be caused by either resins or water-soluble extractives. Spruce knots have been found to contain up to 26% lignans [82] coupled with a low resin content. Pine knots, on the other hand, are known to hold large quantities of resin as well as certain

FIGURE 4.78 After 21 days' storage in a humid environment, the white-painted specimens of some of the tested wood species have suffered substantial discoloration due to extractive bleeding. (Photo reproduced with the kind permission of Peter Böttcher [19].)

FIGURE 4.79 Heartwood extractive bleeding in pine is particularly striking at finger joints.

FIGURE 4.80 The severity of infection by mold fungi on plywood roof soffits can vary considerably (birch plywood was used here). This is most likely due to differences in the nutrient content of individual veneers and thickness variations in the coating.

substances that are also found in pine heartwood (stilbenes) [83]. The differences between the extractives occurring in pine and spruce knots make the prevention of bleed-through difficult. Discoloration often tends to be less pronounced on sunlit than on shaded surfaces.

The water-soluble sugar arabinogalactan may account for up to 25% of some species of larch. At the moment, without special investigations, it is impossible to predict how much arabinogalactan a particular piece of wood contains. Large quantities of arabinogalactan in larch may complicate the bonding process or cause imperfections in coated surfaces. Leached arabinogalactan can also leave sticky brown marks on brightly colored surfaces, which are fortunately easy to remove with water.

The large-scale infection with mold fungi of coated roof soffits made from birch or pine plywoods clearly demonstrates the ability of fungus nutrients to migrate through coatings (Fig. 4.80). Early-morning condensation can cause moisture to accumulate in the exposed surface. As these dry out, the sugars and proteins abundantly present in many sapwoods are transported to the surface [57]. The sheltered location of roof soffits prevents these nutrients from being washed away by rain and they remain on the surface as a food source for mold fungi. The thicker and more water-tight a coating is, the lower the risk of infection. The problem can also be eased by the use of biocidal coatings.

CHAPTER 5

Examples of Failures

However satisfying it is to report on cases for which a clear-cut solution was found, such cases are only part of the truth. The sad reality is that some failure causes can never be established. Several unsolved cases of this kind are examined in Sec. 5.1. The examples in the sections after this focus on several relatively frequent failure causes that tend to manifest themselves in a wide variety of forms. Severe slope of grain, for instance, can lead to deformation, cracks, fractures, and even discolorations.

Sections 5.6–5.10 then take a detailed look at more unusual and complex cases.

5.1 Undetermined Failure Causes

In many cases, a detailed and systematic investigation will bring to light the root cause of failures with reasonable effort. Some failures, however, will remain unsolved or, at best, only a rough assessment of the potential causes will be feasible.

An estimate of the success rate achieved in 600 failure analyses at the Fraunhofer Institute for Wood Research (WKI) in Braunschweig indicated that some 80% of failures were either fully resolved or at least attributed to a highly probable cause. With the remaining 20%, no satisfactory explanation could be found for the failure, even though it was almost always possible to rule out certain factors completely. It is therefore always advisable to inform clients at the very first meeting that not all failure analyses are successful—particularly where the associated effort is expected to remain in reasonable proportion to the financial loss.

5.1.1 Fracture of a Ladder

A ladder made from southern yellow pine fell apart, causing serious injury to the user (Fig. 5.1). One rail of the ladder had broken in two at a height of approximately 30 cm while another rail had only partly fractured. Although the wood of the fully fractured rail was found to have annual rings of irregular width as well as traces of reaction wood, it nonetheless complied with all requirements of the relevant European standards. Moreover, a visual inspection of commercial ladders made from southern yellow pine on sale at DIY stores revealed that wood of this kind is not uncommon. The only partly fractured rail, however, failed to meet European standards due to the presence of compression wood (Fig. 5.2). Yet, further investigations and analysis allowed this clearly substandard rail to be ruled out as the actual cause of the accident.

FIGURE 5.1 One of the rails of a ladder made from southern yellow pine broke apart.

Detailed microscopic examination allowed the presence of compression failures in the wood to be ruled out. (See [66] for a description of compression failure as the cause of fractures in ladder rails.) Nor was there any significant slope of grain. The crack faces of the fully fractured rail were clean, that is, the lack of penetrating dirt argued against the existence of any cracks prior to failure. The rungs had been properly fastened to the rails, that is, the rail had suffered no prior damage during the assembly process. Bending tests performed on sections of all four rails showed roughly the same strengths for each

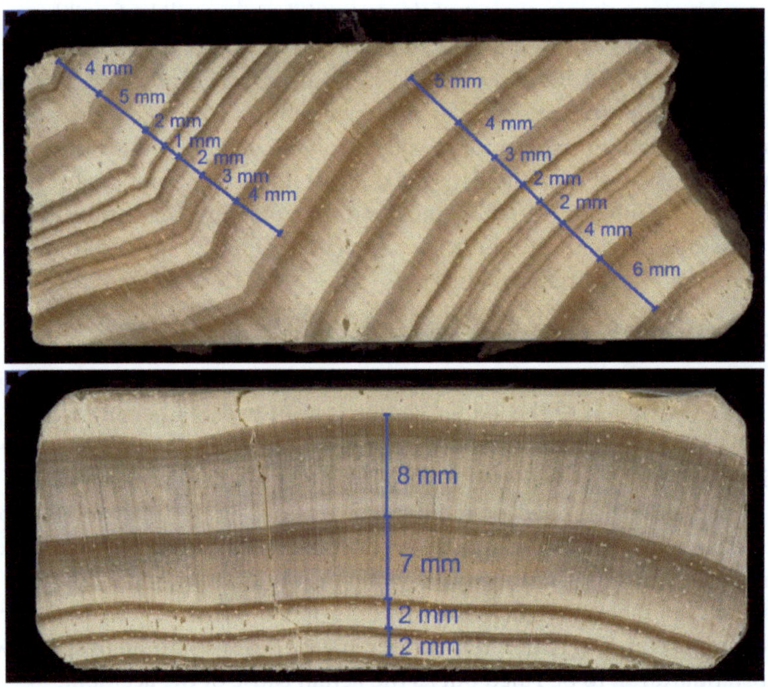

FIGURE 5.2 Polished sections of fully fractured rail (top) and partly fractured rail (bottom).

member, which far exceeded the normal service loads projected for ladder rails. The rails displayed no signs of denting that might have indicated a heavy impact or other form of overstressing prior to the accident.

As the analysis results showed the wood of the fully fractured rail to comply with standards and no fault was found with the ladder construction, this case remains unsolved. Although ladder rails do not normally fracture in the absence of prior damage or dramatic circumstances, no indication could be found of the actual failure cause.

5.1.2 Failure of Spray Adhesive

A spray adhesive was used to bond HPL to plywood. After several months in use as a library countertop, the HPL peeled off over the full surface on both top and rear panel faces (Fig. 5.3). Adhesive was found on all surfaces, though appeared relatively brittle.

Past experience has shown that failures involving spray adhesives are difficult to analyze. Complicating factors for the investigation and assessment include the inevitable variations in adhesive thickness resulting from the manual application process and the highly inhomogeneous bondline at microscopic level. Also, the existing standards and technical guidelines governing required strengths, application quantities, open time and pressure are vague and inadequate. Accordingly, the client was informed at the very first meeting that the chances of finding a clear-cut explanation were slim.

FIGURE 5.3 After three months in service, large areas of the HPL peeled away from the substrate.

The failure distribution, lack of any obvious swelling or shrinkage, and locational factors allowed any overloading due to the environmental conditions be ruled out as a possible cause.

The microscopic examination revealed the presence of adhesive on both plywood and HPL. This indicated that fracturing had occurred within the bondline. Sufficient adhesive was found to be present and the application also appeared relatively uniform for a spray adhesive. These findings pointed to three main possible causes:

1. The adhesive was left "open" for much too long or much too short a period after application. (As the applied product was a solvent-borne contact adhesive, an approximately 10-minute drying time was required before the two adherends were pressed together.)
2. The pressure was too low (possibly in conjunction with too long a drying time for the adhesive).
3. The adhesive was unsuitable or the batch was faulty.

A cartridge of the spray adhesive was obtained and bonding tests were performed with varying parameters. Strength tests on the resulting bonds showed the spray adhesive to be remarkably unaffected by even extreme variations in the process parameters. Both the drastic under-application of adhesive and a massive exceedance of the specified open time had virtually no impact on bonding quality. Bond performance remained acceptable even when the adherends were pressed together only by hand.

Although the IR spectrum of adhesive specimens taken from the failed components differed only in a few details from the adhesive used for the bond tests, these differences were significant. Even 24-hour storage at 103°C of the fresh adhesive from the tests produced no change in the IR spectra, with the adhesive still remaining fairly elastic and slightly tacky. As such high temperatures enormously accelerate chemical changes; it seemed fair to assume that the deviations in the IR spectrum and the brittleness of the failed adhesive could not have been due to gradual processes occurring within a 3-month period.

The following conclusions were drawn:

Either a different (faulty) batch of the same adhesive or a completely different adhesive had been used for the failed countertops. Should the adhesive of another manufacturer have been applied, it is not possible to judge whether or not this was defective as neither the relevant technical data sheet nor the application guidelines could be consulted.

The astonishing resistance displayed by the spray adhesive under diverse application conditions and the client's insistence that only this single type of adhesive had been used made a faulty batch of adhesive the most likely explanation. However, as statements made by clients can never be taken as absolutely reliable, it was not immediately possible to identify the failure cause.

Postscript: A few months later, in connection with another case, we received a specimen with the same failure pattern that had been manufactured at approximately the same time as the product featuring in the above analysis. The same adhesive had apparently been used and this exhibited the same divergent FTIR spectrum as in the previous case. The identical findings from two independent cases very strongly suggested that a defective batch of adhesive was the cause. This knowledge was not, however, available at the time of the original analysis, which is why the original case remained unsolved.

5.1.3 Peeling of Coating on Matched Lumber

The specimen submitted to the Fraunhofer Institute for Wood Research was a tongued-and-grooved board from a roof soffit (Fig. 5.4). The soffit boards had been supplied with a factory-applied priming coat, and the exposed face had received a finishing coat on site. The identity of the used coating materials was unknown. The finish in the

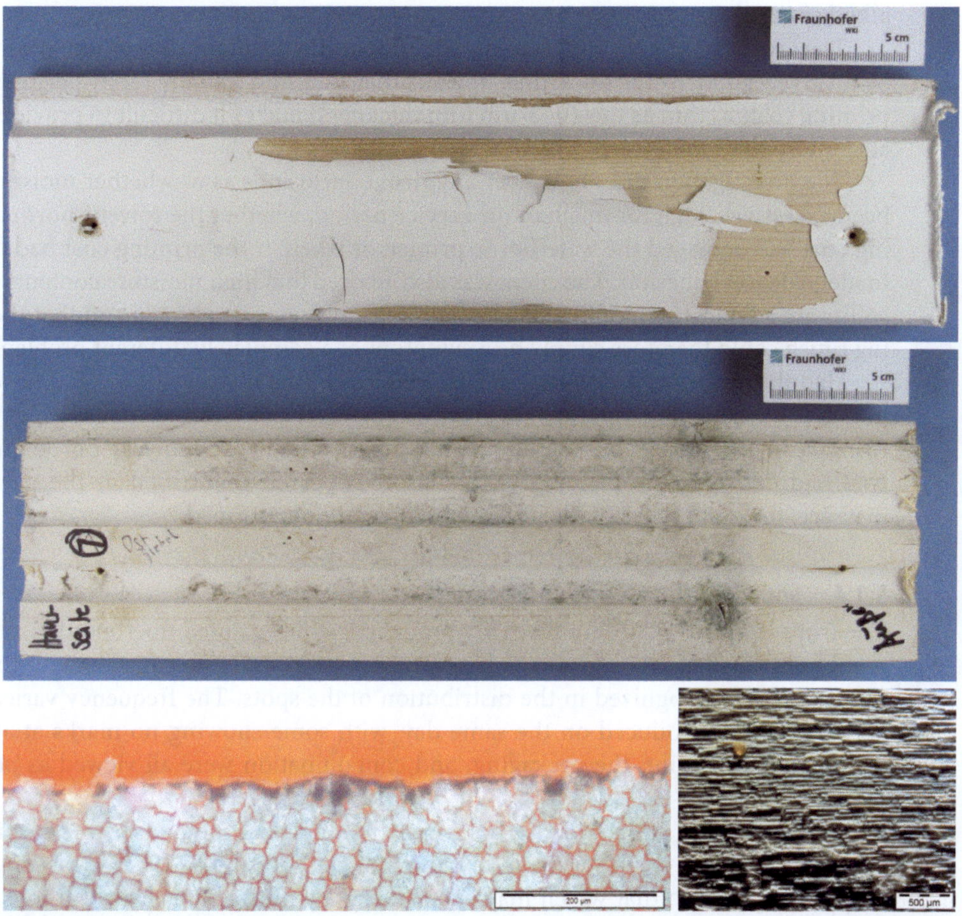

FIGURE 5.4 The top photos show the exposed face of the investigated board with the severely peeled finish and the mold-infected rear face. The bottom micrographs show a cross-section of the priming coat on the rear face of the specimen (left) and the underside of the coating that peeled from the exposed face, with a clear imprint of the individual wood cells (right).

middle section of the exposed face had suffered severe peeling in places while the rear face showed signs of mold. Adhesion was found to be better at the board ends, which had been in contact with the rafters during service. The equivalent zones on the only primed rear face were also free from discoloration by mold. On the basis of this evidence, moisture had clearly been a causal factor. The exact degree of moisture penetration could not, however, be established. Microscopic examination revealed that the loss of adhesion had primarily taken place between the priming coat and the wood. The detailed imprint of the wood observed on the underside of the peeled coating suggested thorough wetting of the substrate by the primer. This allowed an overly viscous priming material or contamination of the wood, for example, by oil or pitch, to be ruled out as a potential cause. IR spectroscopy analysis established that the primer was a waterborne acrylic system while the finishing coat was found to be a solvent-borne alkyd system.

The cross-cut test showed the wet adhesion of the priming coat on the rear face to be relatively good. At the same time, the fact that adhesion tests are rarely performed on priming coats as thin as this (0–3-μm film thickness) makes it difficult to provide a reliable interpretation of the results.

The investigation did not deliver any firm conclusions as to whether moisture had been excessively high for much of the service period, whether the solvent-borne finishing coat had damaged the waterborne primer, or whether the priming coat had offered inadequate wet adhesion. The client was also advised that high moisture contents at roof soffits are a typical sign of inadequate building envelope airtightness and that the peeling finish could therefore simply be a symptom of a completely different problem.

Had this case entailed immense financial losses, the investigation could have been pursued, for example, by performing further chemical analyses and trial coatings on the rear face in order to establish, among other things, whether the solvent-borne finishing coat had damaged the primer. As this effort appeared unjustified in the particular instance, the exact cause of the peeling remained undetermined.

5.1.4 Spots on White Paint

The front and rear faces of industrially produced, white-painted fiberboard displayed a number of small brownish spots only several weeks/months after installation. Initially, no pattern was recognized in the distribution of the spots. The frequency varied even between boards produced on the same day, with some showing no marks at all. The board itself, the paint, the packaging, and contamination were all viewed as possible causes.

Traces of nitrogenous hydrocarbons were detected near the spots by means of TOF-SIMS analysis. The weak spectrum obtained for the spots by FTIR spectroscopy similarly corresponds to that which may occur for many nitrogenous hydrocarbons. FTIR spectroscopy analysis of virtually all the possible substances used in production as well as all types of ambient dust revealed no match with the results obtained for the spots. Ultimately, the chemical analyses were unable to supply any coherent explanation for the spots.

A breakthrough in the failure analysis came with the application of fluorescence techniques using a strong UV torch with a 365-nm wavelength. This showed the visible spots to be only the tip of the iceberg. In reality, the examined surfaces were full of spots, though only the very prominent ones were visible with the naked eye (Fig. 5.5). The now much more numerous spots, including smaller specks, displayed a striking pattern: many of the spots actually comprised one large point accompanied by one or several smaller (secondary) points. The arrangement of points at the spot locations was identical across each individual panel. The number and configuration of secondary points, on the other hand, varied from panel to panel.

This recurrent arrangement provided a useful clue as to how the secondary points arose: The painted surface must have become slightly displaced in relation to an overlying or underlying surface on which a discoloring substance was present. This displacement meant that the discoloring substance left exactly the same pattern of marks at different places on the panel surface. Panels only came into full contact with another surface at two points in the production process—during and after coating. The varying positions of secondary points around the main point on different panels clearly indicated that they can only have arisen during transportation within the factory after coating. The irregular displacement of stacked panels was only possible during

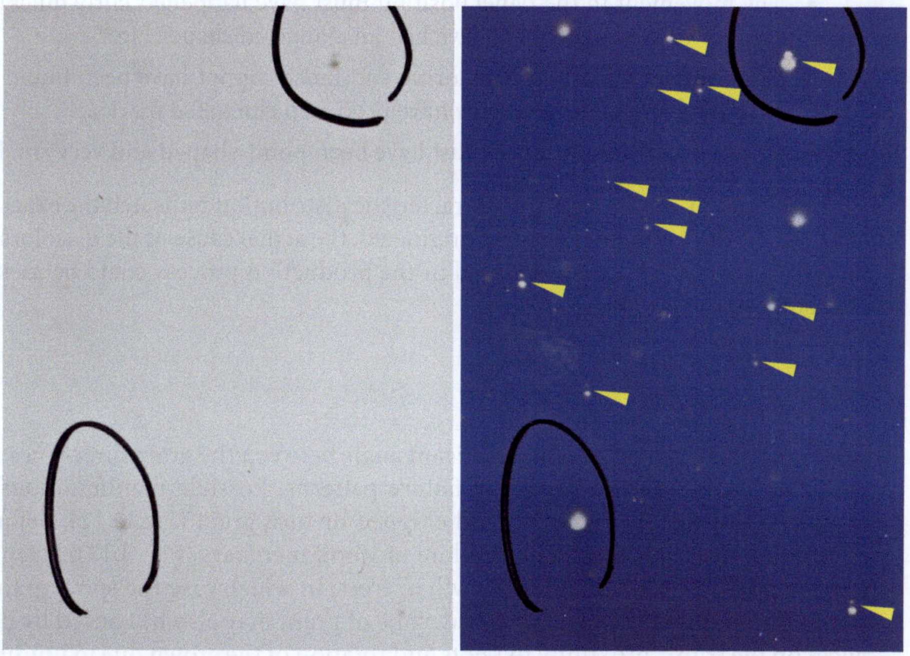

Figure 5.5 Under normal light (left), only the two encircled spots are visible. On account of their fluorescence, many more spots can be seen under strong UV illumination. Particularly striking is the fact that many of the spots actually comprise one larger and one smaller point with the same position and distance relative to each other (arrows).

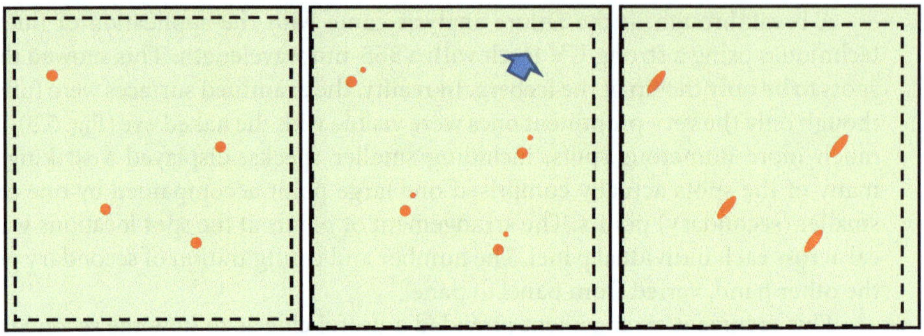

FIGURE 5.6 Diagrammatic explanation of uniform marking patterns on a single panel. If the panels lie on top of each other without movement, the presence of contamination will only cause a single spot. The second spot resulting from movement of the panels in relation to each other will document this relative displacement. Either slow displacement or a contaminant in liquid or paste form would produce elongated marks.

transportation while, in all likelihood, any relative movement during the painting operation could only have been in a single direction.

Further evidence was provided by the appearance of the secondary points:

- The movement in the panel position must have happened fairly quickly or else the marks would probably have had an elongated shape (Fig. 5.6).
- The consistency of the discoloring substance cannot have been liquid or pasty as this, again, would probably have produced elongated marks.
- The discoloring substance must have been point-shaped and very small.

Even though, in this case, the characteristic distribution indicated the exact location at which the discolored patches had originated, the actual cause of the discoloration was not established, as no substance used in the production process could be identified as the source of the problem.

5.2 Slope of Grain

Severe slope of grain, that is, a significant angle between the grain direction and trunk axis, can lead to a wide variety of failure patterns. Possible manifestations include discoloration (Fig. 5.10), cracks and chipped or torn grain (Fig. 5.12), deformation, and the spontaneous mechanical failure of wood members (Fig. 5.13). Steep slope of grain can result from the spiral growth of trees, in which case the spiral grain will be present throughout the log. Localized slope of grain may also be caused by protuberances on the trunk, proximity to knots and rotation of the crown due to unidirectional wind loads.

Slope of grain is sometimes hard to recognize as the grain direction is not always correlated with the observable annual ring structure (Fig. 5.7). Slope of grain is a

FIGURE 5.7 The seasoning checks in both oak posts run along the grain. The grain/crack direction is independent of the orientation of the flat-sawn (tangential) figure.

microstructural phenomenon at the level of wood cells and, hence, a purely visual assessment, particularly on tangential surfaces, may lead to misinterpretation.

The diagrams in Fig. 5.8 show how each new annual ring sits on top of the previous ring like a hollow cone. To the naked eye, all tangential surfaces will show the same flat-sawn figure, regardless of whether the new annual ring was formed with a vertical grain direction or with a significant angle between grain and trunk axis.

In contrast, the grain direction on radial surfaces is always parallel to the visible annual rings. In this case, any slope of grain will always manifest itself as an angle between the grain and the plane of the board (as illustrated in Fig. 5.10).

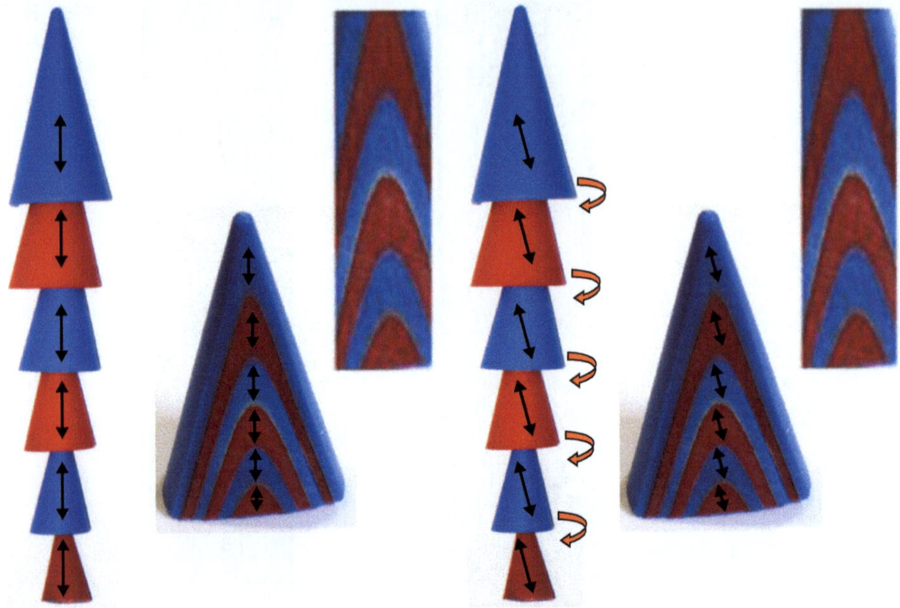

FIGURE 5.8 Diagram showing development and recognizability of spiral grain. Wood can be pictured as a series of hollow cones (annual rings) stacked on top of each other. If each cone is twisted round in relation to the previous one (spiral grain), there is no change in the familiar, visually observable flat-sawn (tangential) figure, even though the wood cells—visible only under the microscope—no longer run in the expected direction.

With larger wood cross-sections, seasoning checks always exactly follow—and thereby indicate—the grain direction. In the absence of seasoning checks, however, sloping grain may go unnoticed. The grain direction in woods with large rays is recognizable with the naked eye as it always corresponds to the orientation of the wood ray cross-sections. If no such structural features are present, the slope of grain can be established by means of microscopic examination, mechanical tests or staining.

On the basis of the procedures described in EN 1310 or AS/NZS 1080.2, a swivel-handled scribe (needle) or similar tool can be used to scratch the slope of grain into the wood surface. To create a better contrast, the resulting depression can then be traced, for example, with a ballpoint pen. Another means of measuring slope of grain involves splitting apart pieces of wood. Here, the observed fracture surface will usually run exactly in the grain direction.

Planed surfaces can be tested using drops of dyed water (ink or cotton blue). As the liquid spreads, its extremities will always run in the grain direction [39] (see Fig. 5.10).

The resin canals in pines often strongly fluoresce under intense UV light, thereby indicating the exact grain direction (see Fig. 3.16).

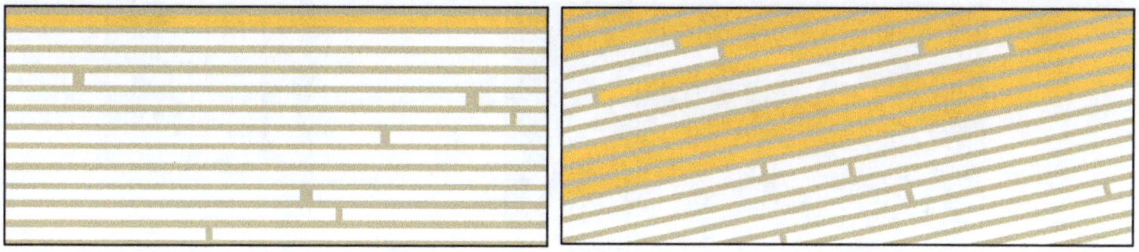

FIGURE 5.9 Schematic illustration of coated side grain with different slopes. The penetration depth of adhesives, stains and other coatings increases as the slope of grain becomes steeper.

As a general rule, the absorption of adhesives, stains and other coatings normally increases as the angle between the wood cells and wood surface widens. Where the cells run parallel to the surface (i.e., 0° angle), penetration into the wood is limited to a depth of around one cell layer (Fig. 5.9). At an angle of only 5°, this already increases to roughly 6 layers [27]. The variations in absorption brought about by the slope of grain are sufficient to cause local discoloration (Fig. 5.10). Other possible causes for similar types of discoloration—should no slope of grain be identified—include application errors or the degradation of pit membranes in the wood due to bacterial attack following prolonged periods of wet storage.

The destruction of pits by bacteria is another possible cause of uneven stain absorption [84]. Pits are tiny "valves" present in great number in every wood cell. Destruction of the central membrane of these valves by bacteria vastly facilitates the penetration of liquids into the cell. Bacterial infection is very rare in normal wood. It may, however, become a problem in wood that has been stored under water or in very wet conditions for a longer period. Pit deterioration can be observed under the microscope (Fig. 5.11).

FIGURE 5.10 The color variations on the surface of a spruce member were caused by the irregular penetration of a transparent coating due to sloping grain. The bottom image shows how the grain direction was made visible by drops of ink on a longitudinal section through the wood.

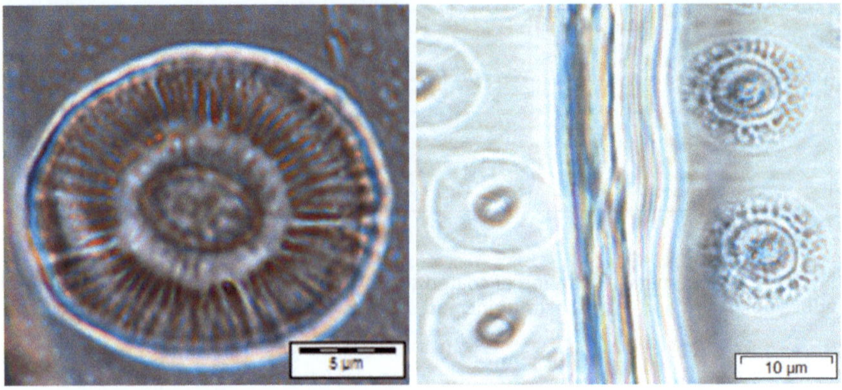

FIGURE 5.11 Left: Sound pit in fir. Right: Sound and bacteria-perforated pits in western red cedar.

Slope of grain can lead to surface chips or tears, especially where the machining speed is too fast and/or the tools are blunt (Fig. 5.12). Apart from their unsightly appearance, such chipped/torn grain can lead to splintering, blistering, bonding failures and discoloration.

Sloping grain can considerably reduce the strength properties of wood (Fig. 5.13), particularly under dynamic loads. Slope of grain as the root cause of fracture in hard maple and white ash baseball bats is convincingly described in [39]. Even a 10° slope of grain may reduce bending strength by 30%. In a project at the University of Graz in

FIGURE 5.12 The edges of plywood strips exhibited a high incidence of splintering. In the left-hand photo, the steep slope of grain is revealed by drops of ink. The polished section on the right shows how the splintering follows the direction of the prominent peeler checks. Here, the splintering resulted from a combination of slope of grain and deep peeler checks.

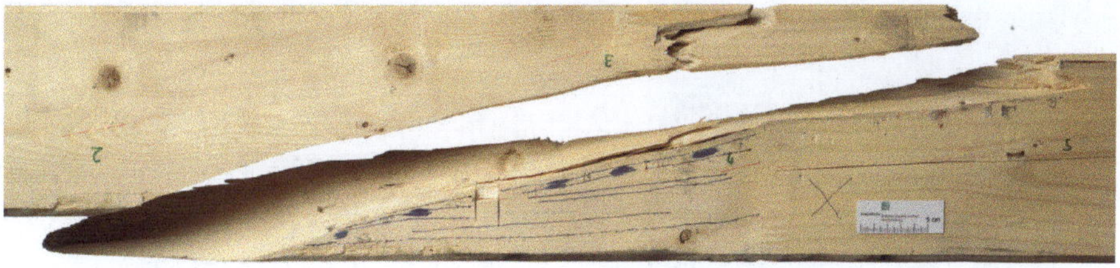

FIGURE 5.13 Fracture in a beam due to unallowable sloping grain of between 10° and 15°. No seasoning checks were present prior to failure to indicate the grain direction. The strength properties of wood, particularly under dynamic loads, can be substantially reduced by steep slope of grain such as that encountered here.

Austria, some 67,000 square-sawn timbers were subjected to mechanical tests under tensile loads of 7 N/mm^2. Of the 133 failures, 34% were attributed to so-called "global" slope of grain (spiral grain, taper) [47]. The report notes that ... *many of the features causing failure would have been difficult, if not impossible, to detect, either visually or by machine, using the standard wood-grading methods currently employed. This conclusion was confirmed by both internal and external experts after thorough assessment of the failures.*

Where both the radial and tangential surfaces show signs of sloping grain, the true compound slope is greater than that on either surface. This compound slope of grain can be determined using AS/NZS 1080.2 or [53].

5.3 Coarse Flakes/Particles/Holes in Surface Layer of Particleboard and Fiberboard

Many characteristics required of wood-based products are clearly laid down by national and international standards. Essentially, however, the properties specified by standards or the manufacturer only apply to the state of the material upon delivery. Difficulties arise where there is reason to suppose that the original characteristics may have been changed by subsequent production processes or service conditions. Other problems may stem from the lack of clear-cut specifications governing the surface texture of wood-based products. Specific cases may therefore call for an assessment as to whether certain surface textures (particle size distribution, presence of foreign matter, holes or bark particles) are still within normal limits. If failures have never occurred with previous deliveries, it may be appropriate to check whether the quality of these deliveries differs significantly from the present quality.

Some particleboard surfaces are dotted with minute holes while others display an almost completely unbroken surface, even under the microscope. Nor is the wettability of wood-based product surfaces identical for all manufacturers. Although variations such as these between the surfaces of different particleboard products may affect coatability, they are not covered by any standards and may also be difficult to measure. In case of failure, the relevant surfaces must be assessed comparatively or with reference to empirical values.

Particleboard and fiberboard may display localized elevations due to coarse particles or foreign matter in the surface layers (Figs. 5.14–5.16). Such particles may "telegraph"

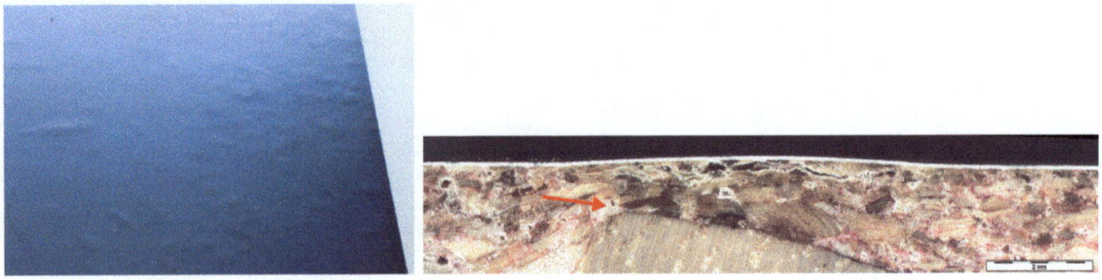

Figure 5.14 The slight elevations in a painted particleboard panel were caused by large, near-surface flakes in the top layer. The cross-section shows a typical large flake and the resulting bulge.

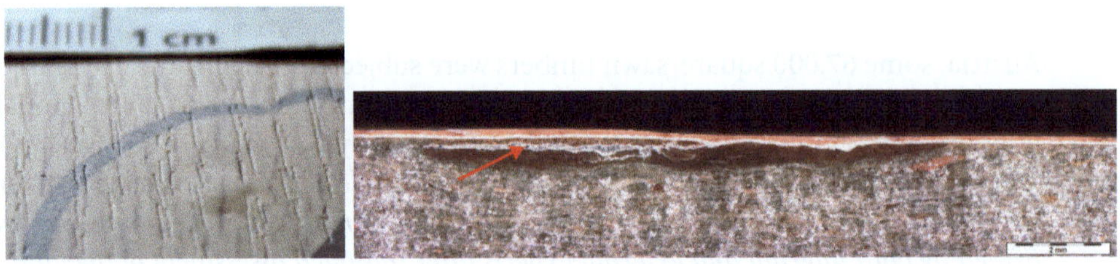

Figure 5.15 The elevations in the coated surface of a fiberboard panel were caused by a large clump of adhesive in the top layer. The cross-section reveals how the hump resulted from a split in the adhesive.

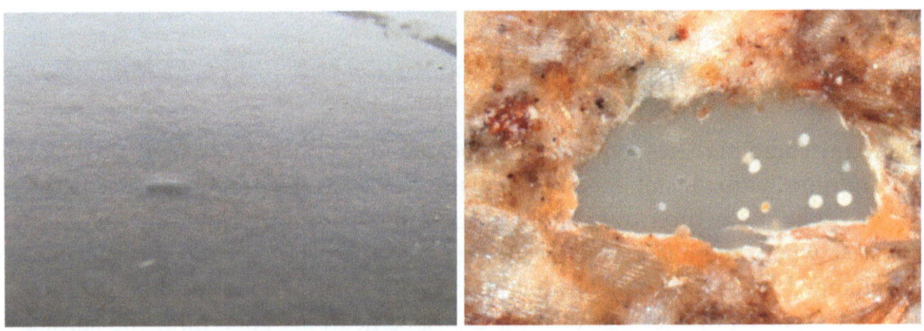

Figure 5.16 Minor elevations were observed on a melamine-coated particleboard panel. To establish their cause, the coating was carefully sanded down over one of the humps. The micrograph of the panel surface under the elevation shows the imperfection to be the result of an entrapped plastic particle.

through subsequently applied coatings to create an uneven finish. With high-gloss finishes, even the tiny elevations or depressions caused by relatively small particles may still be visible through the priming film and coating. Products with larger hollows left by flakes that fall out or are torn out during sanding (Fig. 5.17) are unsuitable for furniture, due to the strict quality standards that apply here.

Wood particles rich in extractives or foreign matter at the surface of particleboard or fiberboard can also lead to discolored patches on film surfaces (Fig. 5.18). The example in Sec. 3.8 (Fig. 3.127) shows how linseeds mixed in among particleboard flakes led to discolorations on backing paper.

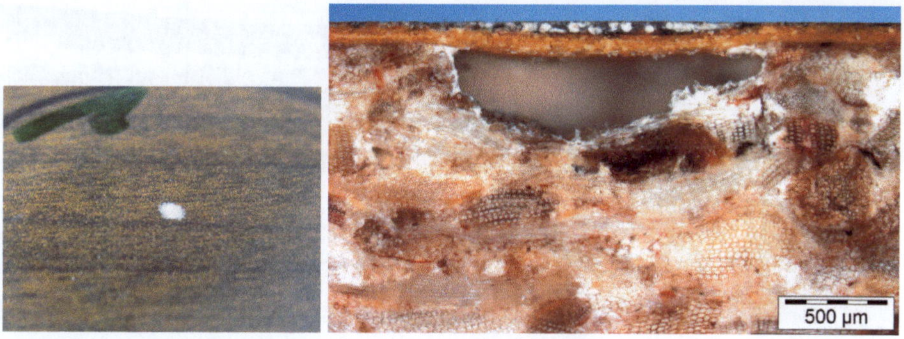

FIGURE 5.17 White spot on melamine paper of particleboard. As the polished section reveals, the discoloration was caused by a hole in the surface layer of the particleboard: here, the light-scattering effect is responsible for the white appearance of the bubbles that formed in the melamine film at this point due to the lack of counterpressure.

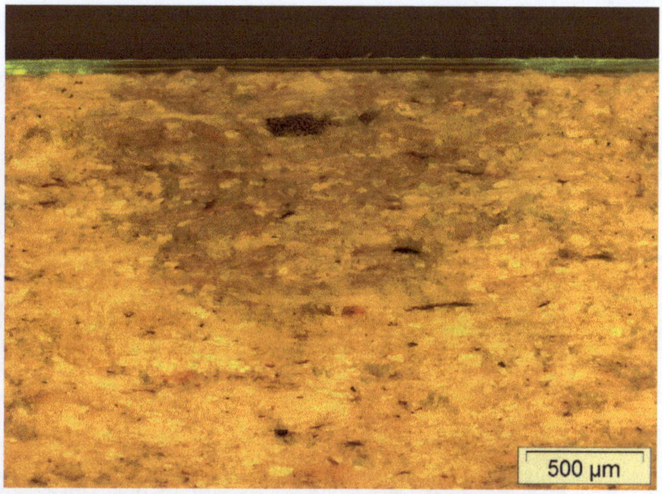

FIGURE 5.18 Polished cross-section of MDF under fluorescence excitation. Extractives from a large particle in an MDF panel can be seen to have migrated through the coating.

5.4 Inadequately Compressed Adhesive and Its Consequences

As a relatively common failure cause, the insufficient compression of adhesive beads has a wide variety of manifestations. During manufacture, adhesive is applied in uniform lines by bead extruders or curtain coaters. Failure to press this into a homogeneous bondline may result in peeling, discoloration, cracks, or swelling. Basically, inadequate adhesive bead compression can have four possible effects:

- Poor adhesion (Fig. 5.19).
- Nonuniform adhesive bleeding that exactly traces the line of the adhesive beads (Fig. 5.20).

FIGURE 5.19 The uncompressed adhesive beads on the right-hand side of the specimen caused the top oak layer of the parquet strip to peel away. The adhesive beads on the left-hand side of the parquet strip are properly compressed. Here, when the specimen is split apart, the top layer fractures in the wood rather than the bondline.

FIGURE 5.20 Light-colored vertical lines on wall paneling in a hotel. The unnatural, linear pattern impairs the appearance of interior finishes of this kind. Typically for this kind of failure, the lines only became noticeable after several years of service. They resulted from whitening in the grain over the areas where adhesive bled through and thus echo the pattern of the adhesive beads.

FIGURE 5.21 Crack in a HPL panel on plywood (left). After the HPL had been sanded down to the bondline and an area of the PVAc adhesive stained with Lugol's iodine (right), the uncompressed adhesive beads became clearly visible.

- Lack of full-surface adhesion: Here, the stresses occurring in the surface layer are not transmitted to the substrate uniformly over the full area, but concentrated on the unbonded zones between the beads. This may encourage crack formation (Fig. 5.21).
- Discontinuous bondline: This means that there is no continuous barrier between the surface layer and base panel to prevent the penetration of liquids and gaseous substances. This may result in local swelling (Sec. 3.1.8, Fig. 3.28).

Possible causes for inadequately compressed adhesive beads include excessively high or low adhesive viscosity, too long a waiting time between adhesive application and pressing, irregular panel thickness and insufficient pressure during adhesive curing.

5.5 Failures due to Iron

Apart from causing discoloration, reactions with iron/ferrous materials can also damage the structure of wood. Moreover, traces of rust may supply important clues about earlier moisture conditions (see Fig. 3.48). Section 3.8 describes a detection method for iron.

Iron salts bring about discoloration in wood through rust formation or reaction with tannins, which then turn a deep black (Fig. 5.22). Intense reactions between tannins and even the minutest quantities of iron have been observed, for example, with the heartwood of oak, black locust, western red cedar, redwood, Douglas fir, and larch.

The discoloring iron particles typically originate from metalworking operations with angle grinders or cleaning with steel wool (Fig. 5.23). When thrown from a short distance by angle grinders, molten particles have been known to fuse with surface finishes and even glass panes. Metal grinding works can thus be identified as the certain source of the failure where fused particles of this kind are encountered. It is at least theoretically possible to determine the precise origin of iron particles using analytical methods. Yet, it would seem that normal structural steels are chemically too similar for any significant differences to be established by EDX elemental analysis.

The wood near iron fasteners is particularly susceptible to degradation, a process often referred to as "wood sickness." Such degradation is accelerated by the presence of

Figure 5.22 Extreme iron discoloration in tatajuba patio decking (photo by Gerald Koch, Thünen Institute of Wood Research). Stain marks that run across the grain are particularly unsightly.

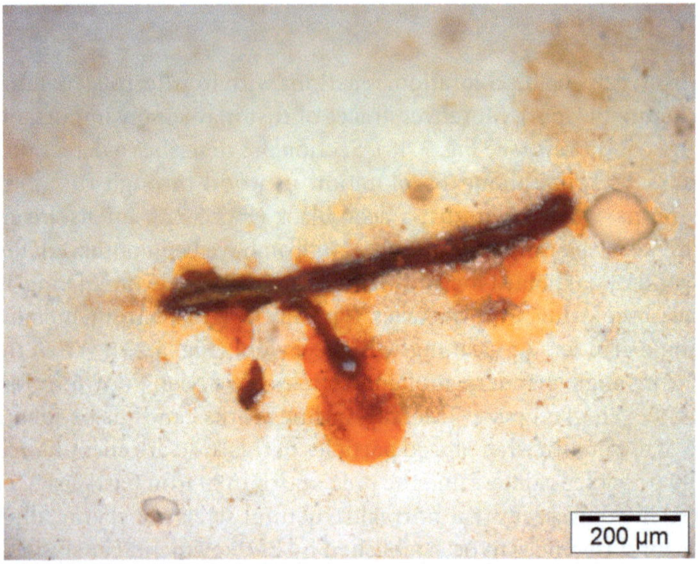

Figure 5.23 The shape of this iron/rust particle suggests steel wool as the origin.

Figure 5.24 Rusting screws can cause severe damage to the wood benching in sauna cabins. Facilities such as this are subject to the combined action of heat, moisture and corrosive chlorides from sweat.

chlorides and elevated temperatures (Fig. 5.24). Shifts in pH values and the catalytic effect of iron ions are the probable causes of this deterioration [16]. Acting in combination with hydrochloric acid vapors, iron salts can cause severe damage to wood (Fig. 5.25).

Given that corroding iron expands to around four times its original volume, large rusting bolts can also induce the mechanical failure of wood members (Fig. 5.26).

Figure 5.25 Iron chlorides forming from hydrochloric acid vapors in a hot-dip galvanizing plant had a devastating impact on the surrounding wood. In its final state, the wood assumed a tar-like color and consistency.

FIGURE 5.26 The increase in volume of rusting bolts caused this beam to split.

5.6 Acid Damage to HPL

Surfaces can be destroyed through the use of unsuitable cleaning agents. The strong acids present in some cleaners will—if not properly diluted in accordance with the instructions—attack melamine surfaces and other finish types. Degraded melamine surfaces are typically found around sinks, faucets and valves. It is likely, here, that the acidic agents causing the damage are used to clean the metal fittings. Any cleaning agent trapped in the capillaries between countertop and metal fittings, if not removed when the surface is washed down with water, will proceed to destroy the countertop finish. To the naked eye, the chemical degradation of melamine surfaces takes the form of discoloration or incrustation. Only under the microscope does the deterioration of the finish become apparent (Fig. 5.27).

In the featured example, it was still necessary to check whether a compatible/properly diluted cleaner had been used or whether the HPL surface itself was defective due to inadequate resistance to standard cleaning products.

An initial test was performed to establish whether the acidic cleaner, in the concentration specified by the manufacturer, is at all able to cause the observed failures. Only the undiluted product was found to induce comparable failures (Fig. 5.28). In properly diluted form, the cleaner brought about no change in the HPL surface even after 6 days' storage at 50°C. To check whether the HPL was nonetheless abnormally sensitive to acids, two other commercial HPL products manufactured by other companies were also tested. When exposed to the undiluted cleaner, all materials suffered the same deterioration.

The conducted tests by no means prove conclusively that undiluted cleaning agent was used. They merely show this to be a very plausible scenario and suggest that, used

FIGURE 5.27 Top view and microscopic cross-section of melamine surface at edge of a sink, destroyed by concentrated acid. The cross-section shows how the acid has not only attacked the surface but has even caused splits in the HPL.

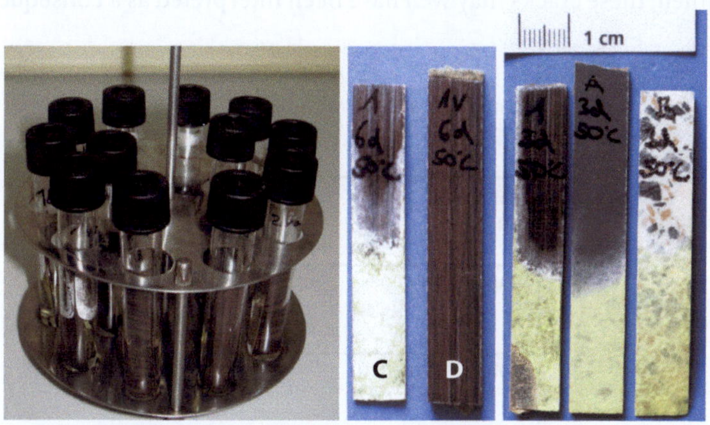

FIGURE 5.28 The HPL from the failure was stored in concentrated (C) and diluted (D) acidic cleaners (middle) and together with two other commercial HPL products in the concentrated solution (right).

in the prescribed concentration, the cleaning agent would have been very unlikely to cause such a failure. If necessary, it would also be possible to confirm the probable scenario by chemical analyses.

5.7 Premature Wear of Coating

After installing a beech stairway, a finish carpenter had applied a two-component polyurethane coating. The client complained of heavy signs of wear that appeared after only a short period of use. In order to establish the cause, the Fraunhofer Institute

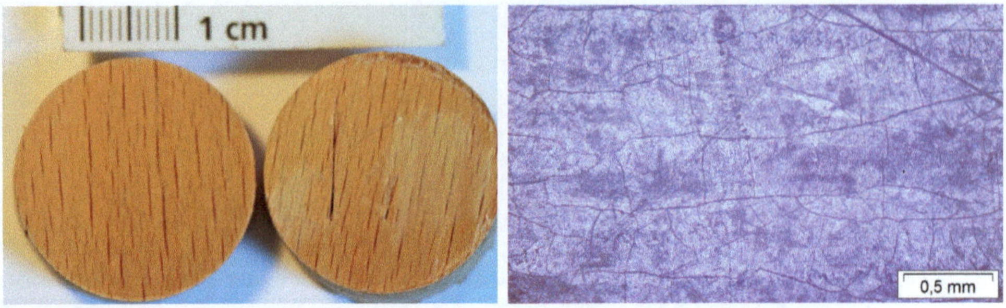

FIGURE 5.29 Drill cores from a stairway (left). Under the microscope, numerous fine coating cracks on the seemingly intact drill core revealed the finish itself to be the source of the problem (right). This allowed factors such as harsh service conditions to be ruled out as a possible cause.

for Wood Research requested the submission of two drill cores: One from the damaged area and one without any visible damage from outside of the foot-trafficked stair surface. As often happens, the apparently intact specimen proved the more helpful in clarifying the failure mechanism. Viewed under the microscope, the surface of this specimen exhibited a large number of typical coating cracks (Fig. 5.29). On the failed specimen, these cracks may well have been interpreted as a consequence of the flaking

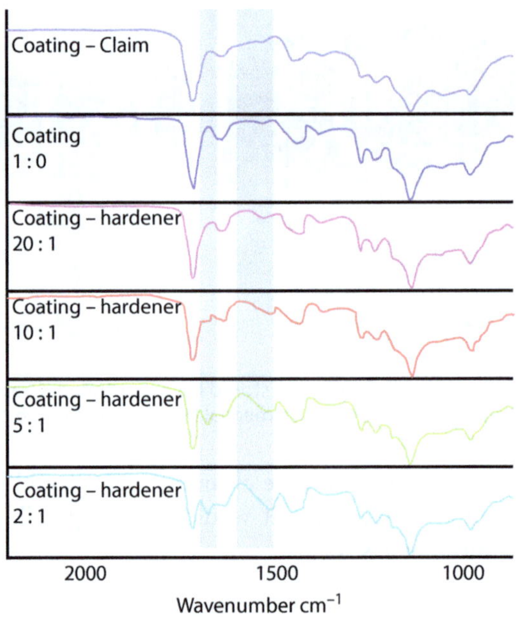

FIGURE 5.30 IR spectra for various proportions of coating and hardener, compared to coating from failure (top spectrum). The 1,686-cm^{-1} and 1,512-cm^{-1} wavenumber bands were used to estimate the proportion of hardener in the prepared finish. The IR spectra prove that the failed coating contained no or virtually no hardener.

process. Due to their presence on the "undamaged" specimen, however, the cracks clearly indicated a coating failure. While an incompatible chemical treatment could conceivably have damaged the coating, a faulty batch of coating material or the improper proportioning of the two finish components offered a more plausible explanation.

To check whether incorrect proportioning of the hardener could be responsible for the premature failure of the two-component finish, fresh material was obtained from the manufacturer and mixed in various ratios. On the basis of absorption at wavenumbers 1,686 cm^{-1} and 1,512 cm^{-1}, the IR spectroscopy analysis revealed that the coating from the failure best matched the coating without hardener (Fig. 5.30). The hardener had evidently been forgotten or added in an extremely low proportion.

5.8 Did Interior Environmental Conditions Cause Blistering?

Birch plywood panels with maple face veneer were installed to enclose the riding area in an indoor arena. After several months, debonding of the face veneer caused blisters to appear on many of the panels. The plywood panel supplier claimed that the interior environmental conditions had been responsible for the failure.

The specimens sent for analysis confirmed the failure distribution that was already apparent in the photos of the installed panels: the blisters always occurred within the center of the panels and hardly ever near the edges. This finding alone more or less allowed unsuitable environmental conditions to be eliminated as a potential failure cause. This is because, normally, the impact of adverse environments is greatest at the panel edge, where peeling will occur.

Staining of the underside of the detached veneer and underlying panel with Lugol's iodine brought three important facts to light (Fig. 5.31):

1. A PVAc-based adhesive was used to bond the product, as only this adhesive type turns reddish-brown in the presence of Lugol's iodine.
2. As adhesive was clearly present on the panel, sufficient adhesive had been applied.
3. No adhesive was present on the underside of the detached veneer. This is either because the adhesive failed to bond to the veneer or because the adhesive never came into contact with the veneer.

When viewed under the microscope, the adhesive surface mainly displayed the texture typical of an adhesive that cures without contact to the overlying veneer (Fig. 5.32). This finding showed the cause of the blisters to stem from the panel manufacturing process. At those locations where the veneer had later debonded, there had been no contact between adhesive and veneer during curing.

The cross-section shows the veneer to be approximately 0.1 mm thinner in the debonded areas than above the well-adhering zones (Fig. 5.33). It is possible to rule out the eventuality that the veneer was wet and swollen at these locations during sanding, before shrinking heavily in relation to the surrounding dry veneer: this is because the

Figure 5.31 Lugol's iodine was applied to the underside of the detached veneer and the underlying surface of the plywood. The initially colorless PVAc adhesive stained reddish-brown.

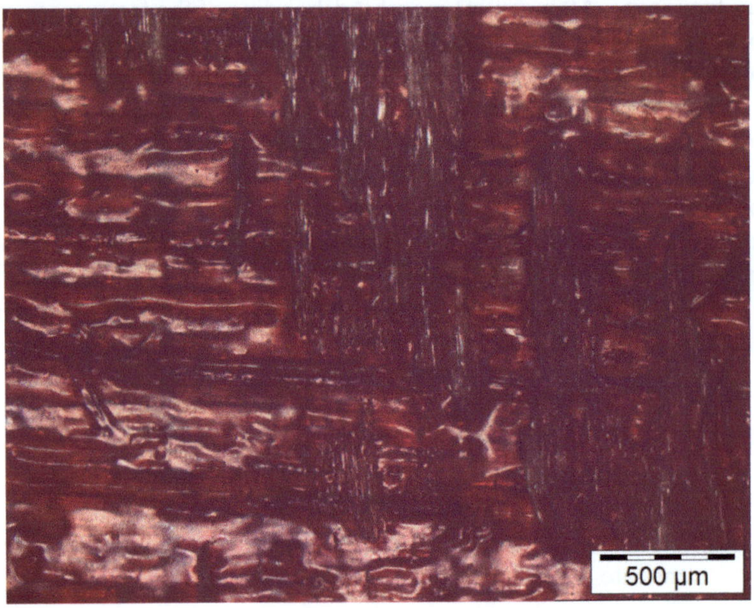

Figure 5.32 Staining of the adhesive with Lugol's iodine reveals the almost complete lack of contact between adhesive and adherend at the blister locations.

maximum possible radial shrinkage for 0.47-mm-thick maple veneer would only have been around 0.02 mm. Hence, the veneer in the debonded zones is very unlikely to have been in contact with the substrate even before sanding. The thinner veneer at the blister locations thus confirms the assumption that the bonding failures occurred before the panels were coated. The fault had resulted from the process adopted to bond the veneers. Adverse environmental conditions can be ruled out as a possible cause of the blisters.

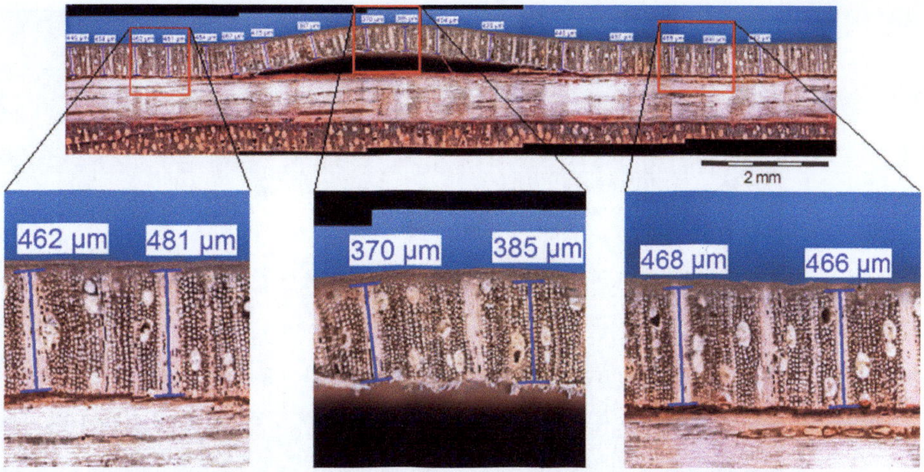

FIGURE 5.33 This cross-section through a blister in veneered plywood was assembled from five separate images. The veneer in the debonded area was found to be around 0.1 mm thinner than in the well-adhering zones. In this particular case, the measurements are likely to be fairly reliable as the wood rays run exactly perpendicular to the face of the (rotary-cut) veneer. Any mechanical stress (crushing) or deformation suffered by the veneer would have been visible in the wood rays. It can thus be concluded that the variations in veneer thickness already existed prior to coating.

5.9 Diagonal Grain Orientation as Cause of Parquet Failure

A premium-quality, custom-designed parquet floor suffered severe damage over an area of several hundred square meters during the first heating season after installation. Some of the corners of the square panels had curled up by several millimeters (Fig. 5.34). The floor finish comprised three-layer parquet with a 7-mm-thick, smoked-oak top layer. It was bonded to a heated screed and the surface was oiled.

During the on-site inspection, the curled corners were only observed on flat-grained top layers. Moreover, given that only corners on the diagonal perpendicular to the grain were affected, it was fair to assume that shrinkage of the top layer was a causal factor. Nonetheless, no other moisture-induced damage, for example, in the form of excessively wide joints or conspicuous checks, was encountered.

With all panel specimens taken during the site visit, the same failure pattern was found at the raised corners: here, the panel core had split at the groove base (Fig. 5.35).

All the findings suggested that the top layer had shrunk during service and that the resulting stresses had led to the split in the core.

The following scenarios (and, of course, combinations thereof) were thought to provide a satisfactory explanation of the observed phenomenon:

1. An excessively high floor heating temperature had caused overdrying of the parquet, such that failures were inevitable.
2. The moisture content of the top layer was much too high when the parquet floor was laid.
3. The parquet flooring was subject to design flaws or material-related weaknesses.

Figure 5.34 Three-layer parquet with smoked-oak top layer, bonded to a heated screed and with oiled surface. The grain is diagonal to the panel edges. The direction of maximum swelling and shrinkage, which is perpendicular to this, thus runs exactly between two of the corners. The site visit took place in summer. The users reported that the corner curling was much more pronounced in the heating season.

Figure 5.35 The curling at the panel corners had induced a split in the core, which originated at the angle in the groove base.

No measurements were available to establish the moisture content of the parquet during service. According to the users, the specified maximum temperature of the floor heating system was never exceeded. The parquet manufacturer, on the other hand, claimed that the allowable temperature had been substantially exceeded even during commissioning of the heating installation. A strong argument against any severe overdrying of the top layers during service was the absence of the surface checking typically associated with this. With flat-grained lumber, at least fine checks would have been expected. It was nonetheless unfeasible to reconstruct the previous environmental conditions. The absence of checks in the parquet and of drying-related failures on the other wood assemblies in the building did, however, suggest that extreme overdrying was improbable.

The moisture content of the top layer at the time of manufacture was estimated by cutting this away from the panel core: the sizes of the cut-free layers were found to be roughly identical when measured under standard environmental conditions of 20°C/50% relative humidity (RH). This indicates that the top layer cannot have had an excessively high-moisture content at the crucial time of bonding to the panel core.

At first sight, the design of the parquet flooring in no way appeared out of the ordinary. Each panel had an 8.5-mm-deep groove on all four sides and was connected to the surrounding frame by a loose tongue. The only conspicuous feature was the unusual grain direction of the top layer, whereby the diagonal between two of the corners was perpendicular to the grain. With this arrangement, the loads acting on the grooves at these corners evidently differ from standard designs, where the grain runs parallel to the panel edges. In the latter case, the stresses acting on the grooves during shrinkage are uniformly distributed over the entire panel length. In the investigated case, however, these stresses were concentrated at a single point in the groove. This was the exact position where the splits occurred. A series of tests were performed to determine the implications of the diagonal grain orientation.

Similar parquet panels measuring 90 mm × 90 mm with different grain directions and groove sizes/shapes were bonded to tiles (Fig. 5.36) and exposed to various environmental conditions (20°C/22% RH, 20°C/15% RH, and 40°C/11% RH). The installed

Figure 5.36 Top views and cross-sections showing grain directions and groove geometries (0 mm, 4-mm and 8-mm rectangular groove, 9-mm groove with rounded base) of tested specimens. The tiles to which the specimens were bonded prior to the environmental tests can be seen on the left-hand side of each cross-section.

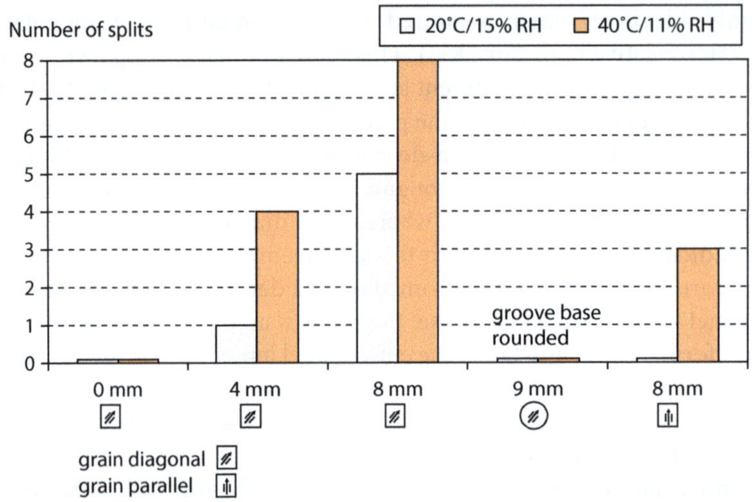

FIGURE 5.37 Number of splits in groove base after two of the three environmental tests. None of the specimens suffered damage at 20°C/22% RH.

floor heating system and reduced humidity directly over the warm parquet surface mean that at least the 22% RH and 15% RH test conditions are not unrealistic.

Storage of the specimens under different environmental conditions clearly revealed the significance of the various parameters with regard to splits in the core (Fig. 5.37).

- The specimens with grain running parallel to the panel edges suffered no splitting at 15% RH, even with an 8-mm-deep groove.
- The specimens with diagonal grain orientation suffered a greater number of splits as the groove depth increased. With an 8-mm-deep groove, splits were observed at five angles in the groove base after storage at 15% RH.
- The specimens with diagonal grain direction and a rounded, 9-mm-deep groove suffered no splitting, even at 11% RH.

The conclusion drawn from these tests is that the unnecessarily deep (8.5 mm) and rectangular groove with loose tongues significantly increased the sensitivity of this parquet to dry environmental conditions. The overlooked consequence of the diagonal grain direction was a concentration of all shrinkage stresses precisely at the upper edge of the angle in the groove base, that is, theoretically at a single point (Fig. 5.38). The simple provision of a rounded groove base, a reduction of the groove depth or even two drops of adhesive on the loose tongue at the corner would have spread these stresses more uniformly and protected the parquet floor against splitting.

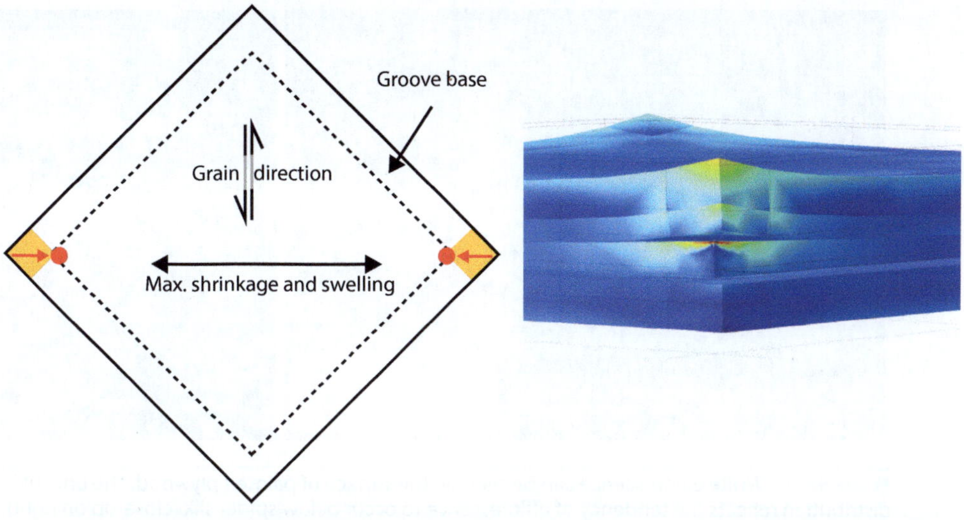

FIGURE 5.38 The diagonal grain direction of the oak created stress peaks at the angles in the groove base. Stress calculations for the parquet construction (right) confirmed the stress peaks caused by the shrinkage of the top layer (courtesy of Pablo Guindos).

I always view this case with a good deal of humility. Had the parquet contractor sent me the construction drawings for checking prior to installation, I would probably not have recognized the problems that now appear so obvious. Failure investigators are always in a privileged situation in that they only step in after failures have occurred, their only job being to hunt down the cause. In doing so, they often gain insights that are unavailable to the contractors entrusted with realizing innovative designs.

5.10 Unusual Distribution of Efflorescence

A gray-painted, rain-sheltered plywood facade exhibited a strange pattern of white patches that reappeared after cleaning. Examination with a reflected-light microscope revealed that the unusual distribution corresponded to the occurrence of spider silk on the surface (Fig. 5.39). Large quantities of spider silk were present on the facade and a white substance was observed to have accumulated below and on the silk. The phenomenon was similar to that occasionally witnessed in the open country in frosty weather, whereby normally invisible spider webs attract a covering of condensing hoar frost that makes them visible against a color background.

The specimen cross-section revealed the presence of white crystalline structures, resembling salts, in the plywood adhesive (Fig. 5.40). As these types of structure do not occur in normal adhesives, it was thought that they might be identical to the deposits on the panel surface. This possibility was investigated by means of EDX spectroscopy analysis, which permits the exact identification of the elemental composition of salts.

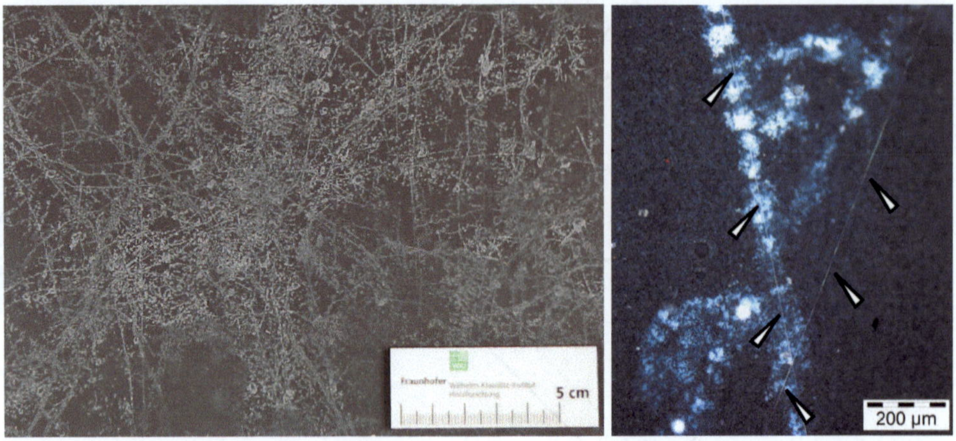

Figure 5.39 White efflorescence can be seen on the surface of painted plywood. The unusual distribution reflects the tendency of efflorescence to occur below spider silk (close-up on right).

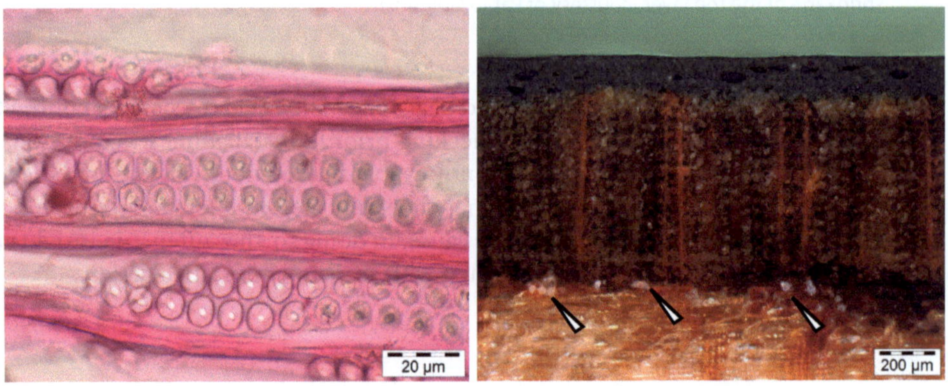

Figure 5.40 The plywood species was identified as araucaria by the typical arrangement of pits (left). The cross-section reveals the presence of white crystals in the bondline (right).

The EDX analysis showed both the deposits on the surface and the white particles in the bondline to be a compound containing sulfur, sodium and oxygen, for example, sodium sulfate or sodium sulfite (Fig. 5.41). The plywood was made from a species of araucaria, suggesting South America as its origin.

Although the EDX analysis showed the adhesive to contain unusual, water-soluble substances, this was by no means the final step in the failure investigation. The visible defect depended on the substances in the adhesive being dissolved and transported to the panel surface by water. In this particular case, the distribution of the efflorescence and traces of mold fungi on the rear panel face showed the absence of a vapor barrier to have resulted in water transport to the surface. The moisture condensing on the rear panel face had penetrated through the plywood and carried the efflorescent salts to the

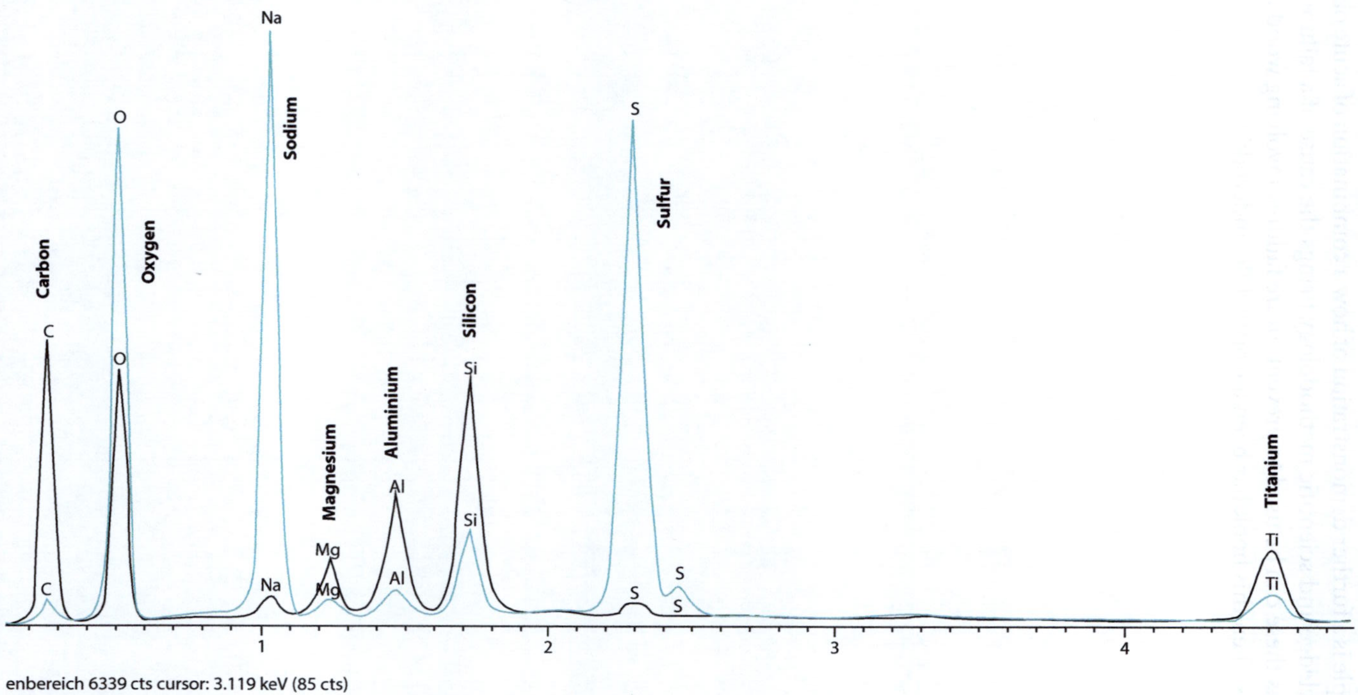

FIGURE 5.41 EDX spectra of white deposits (light-blue line) and paint (black line). The white deposits primarily consist of sulfur, sodium and oxygen. As expected, the paint comprises carbon, oxygen, silicon, aluminum, magnesium and titanium.

painted outdoor face. Consequently, the efflorescence was merely a symptom of inadequate moisture control in the building envelope.

This example is a further demonstration of how a combination of acute observation, practical knowledge and scientific methodology brings the cause of a failure to light. If insights such as these can be used to prevent future failures involving wood and wood-based products, then this book has been a worthwhile undertaking.

Appendix

A.1 List of Abbreviations and Conversion of Units

°C	Degree Celsius Degree Fahrenheit = °C × 1.8 + 32	HPL	High-pressure laminate
m	Meter = 39.4 in.	IR	Infrared (spectroscopy)
cm	Centimeter = 1/100 m 1 cm = 0.394 in. 1 in. = 2.54 cm	ISO	International Organization for Standardization
mm	Millimeter = 1/1,000 m 25.4 mm = 1 in.	LED	Light-emitting diode
µm	Micrometer = 1/1,000 mm	LPL	Low-pressure laminate = MFB
nm	Nanometer = 1/1000 µm	MDF	Medium-density fiberboard
g	Gram = 0.0353 oz	MF	Melamine-formaldehyde (resin)
kg	Kilogram ≈ 2.205 lb	MFB	Melamine-faced board
mL	Milliliter ≈ 0.061 $in.^3$	MUF	Melamine-urea-formaldehyde (resin)
AS/NZS	Australian/New Zealand Standard	PF	Phenol-formaldehyde (resin)
ASTM	ASTM International (formerly *American Society for Testing and Materials)*	PU	Polyurethane
CPL	Continuous-pressure laminate	PVAc	Polyvinyl acetate
DACA	Dimethylaminocinnamaldehyde	RH	Relative humidity
DPL	Direct-pressure laminate	SANS	South African National Standard
DSC	Differential scanning calorimetry	SEM	Scanning electron microscope
EDX	Energy-dispersive X-ray (spectroscopy)	TOF/SIMS	Time-of-flight secondary ion mass spectroscopy
EN	European standard	UF	Urea-formaldehyde (resin)
FTIR	Fourier-transform-infrared (spectroscopy)	USB	Universal serial bus
GC/MS	Gas chromatography/mass spectrometry	UV	Ultraviolet (radiation)

A.2 Some Comments on Acceptable Humidity Levels

Every spring, the Fraunhofer Institute for Wood Research (WKI) in Braunschweig investigates cases of cracking in materials that occurred on cold winter days, when the relative humidity inside heated and partially ventilated buildings was extremely low. At the same time of year, the WKI also deals with cases of cracking, discoloration and other phenomena occurring in winter in unventilated buildings with very high relative humidity caused by freshly installed concrete or plasterwork.

In my experience, many investigators are only too glad to seize on any evidence of relative humidity that they consider too high or too low. Humidity can then be identified as the failure cause and the case closed. But what ranges of relative humidity should a particular product be able to accommodate? What if the product was clearly of unacceptable quality *and* the humidity was beyond the specified limits? And why is it that the vast majority of wood and wood-based products experience no major problems when exposed to above-average or below-average humidity? What distinguishes the products that fail from those that suffer no severe problems?

In the German-language version of this book, I included some comments on acceptable humidity in buildings. In Germany, there are no regulations governing the humidity or dryness levels to be accommodated, without failure, by furniture, flooring, and other wood or wood-based products. The following passage from the German-language edition is applicable only to the climatic conditions in Germany and is no more than my own personal opinion.

The swelling and shrinkage of wood can lead to a wide variety of failure types. A question that often arises in failure analysis is whether the material (or assembly) has reacted oversensitively to still-acceptable environmental conditions or whether these conditions imposed moisture loads so far beyond the tolerable range for the material that the conditions (and thus normally the user) can be made wholly responsible for the failure. It is often difficult to reach a clear conclusion on this issue due to the generally inadequate definition of allowable service conditions and the basic difficulty of reconstructing effective humid conditions.

However, on the basis of natural climatic conditions and the internal environment required for human comfort, the following limits can be postulated for components installed in living spaces:

A.2.1 Upper Humidity Limits in Summer Months

Permanently open windows can, in principle, be regarded as normal and allowable in late spring, summer and early fall. Accordingly, all materials and components should be able to accommodate the outdoor climatic conditions in summer and the mid-season months without failure. In this context, failure-free performance does not imply strict compliance with the ideal condition. Any deviations that occur under these environmental conditions, however, must not lead to lasting imperfections or in any way limit serviceability. Hence, neither a slight telegraphing of concealed edge bands during humid summer periods nor the largely reversible opening of joints in parquet floors in winter should be classed as failures in this sense.

In late spring, summer, and early fall, the average relative humidity in Germany/central Europe can reach 75% or more and persist for a period of several days or weeks. This means that all materials should be able to withstand a minimum relative humidity of 75% for at least several days without failure.

A.2.2 Lower Humidity Limits in Winter Months

The situation in the winter months differs from that in summer and parts of the spring/fall period in that user behavior and technical factors such as window airtightness or humidification systems can then have a major impact on ambient relative humidity. In winter, the water content of the outside air ranges between 2 g and 3 g/kg dry air, equivalent to a relative humidity of approximately 60% at 0°C. As soon as this air is heated to a temperature of 20°C, the relative humidity drops to less than 20%. Low humidity such as this is particularly common in older buildings, where the windows are not properly sealed.

EN 13779 specifies 30% as the allowable lower limit for relative humidity. This standard also states the following: "*The lower limit of 30% is given to prevent dry eyes and mucous membrane irritation. However in severe climates it is acceptable to have lower humidity for a limited period, to be agreed between client and designer and taking into account local norms and preferences. ...*" Hence, even relative humidities of 25% or so are not outside of the range of acceptable conditions for a limited time during very cold winter periods.

A.2.3 Proposal for Evaluation of Upper and Lower Humidity Limits

There is no clear-cut definition of the maximum and minimum relative humidity at which construction materials and assemblies are required to perform without failure.

One way of approaching this issue is to initially consider all permanent, humidity-related failure features on sensitive components that have been verifiably exposed to relative humidity below 25% or above 75% as being attributable to the interior environmental conditions, and thus within the scope of the user's responsibility. "Sensitive" components may be deemed to include elements such as internal doors, furniture and parquet flooring. By the same token, all assemblies and components which can be presumed to be installed in the building prior to the plastering and cementitious topping works (e.g., external doors, windows, wall paneling) should be able to withstand higher humidity without failure. Here, a relative humidity of 85% can be taken as an appropriate upper limit. It is possible to estimate the point in time at which conditions may become critical for a particular component on the basis of the preceding period needed to achieve moisture equilibrium. This is because adverse environmental conditions only become critical as of the time the endangered components approach moisture equilibrium. With an obeche picture frame, for instance, this situation may be reached within a day. For parquet flooring, the change in moisture content can be assumed to take much longer. Rapp et al. [72] describes how the moisture content of beech parquet changes by 2% in 18 days in response to an adjustment of environmental conditions. At 52 days, oak parquet needs almost three times as long.

A.2.4 Other Causes Despite Unallowable Service Conditions

Even where humidity is outside of the stated limits, the components must still be investigated for possible defects. Any existing imperfections that only become apparent due to the adverse service conditions must at least be regarded as a contributory cause of the failure. This is justified by the fact that the component would have been able to withstand the critical environmental conditions had the defect not existed.

A.3 Scientific Names of Featured Wood Species and Organisms

Abura	*Hallea stipulosa*	Meranti	*Shorea spp.*
Araucaria	*Araucaria spp.*	Merbau	*Intsia spp.*
Ash	*Fraxinus spp.*	Niangon	*Herietia spp.*
Beech	*Fagus sylvatica*	Oak	*Quercus spp.*
Birch	*Betula spp.*	Obeche	*Triplochiton scleroxylon*
Black locust	*Robinia pseudoaccacia*	Spruce	*Picea abies*
Brown terminalia	*Terminalia spp.*	Southern yellow pine	*Pinus spp.*
Cherry	*Prunus spp.*	Sapele	*Entandophragma cylindricum*
Douglas fir	*Pseudotsuga menziesii*	Scots pine	*Pinus sylvestris*
East Indian rosewood	*Dalbergia spp.*	Tatajuba	*Bagassa guianensis*
Fir	*Abies alba*	Teak	*Tectona grandis*
Honey locust	*Gletitsia triacanthos*	Wenge	*Millettia laurentii*
Iroko	*Chlorophora excelsa*	Western red cedar	*Thuja plicata*
Jatoba	*Hymenaea courbaril*	**Organisms**	
Kotibe	*Nesogardonia spp.*	Cellar fungus	*Coniophora puteana*
Larch	*Larix spp.*	Deathwatch beetle	*Xestobium ruffovillosum*
Mahogany	*Swietenia macrophylla*	Longhorn beetle	*Hylotroupes bajalus*
Mansonia	*Mansonia altissima*	Lyctus powder-post beetle	*Lyctus brunneus*
Maple	*Acer spp.*		

A.4 CAS Numbers of Chemicals

Astra blue	CAS No. 82864-57-1	Iron (III) chloride	CAS No. 10025-77-1
Cotton blue	CAS No. 28631-66-5	Lugol's iodine	CAS No. 12298-68-9
Dimethylamino-cinnamaldehyde	CAS No. 6203-18-5	Nile Blue Sulfate	CAS No. 3625-57-8
Ethanol	CAS No. 64-17-5	Ninhydrin	CAS No. 485-47-2
Fast Blue B salt	CAS No. 119-90-4	Potassium thiocyanate	CAS No. 333-20-0
Fast Red B salt	CAS No. 56315-29-8	Povidone-iodine	CAS No. 25655-41-8
Hydrochloric acid	CAS No. 7647-01-0	Safranin	CAS No. 477-73-6
Hydrogen peroxide	CAS No. 7722-84-1	Sudan IV	CAS No. 85-83-6
Iodine	CAS No. 7553-56-2	Variamine Blue RT salt	CAS No. 4477-28-5

References

[1] AS/NZS 1080.1:2012: Timber—Methods of test. Method 1: Moisture content. Standards Australia International, Sydney.

[2] AS/NZS 1080.2:2006: Timber—Methods of test. Method 2: Slope of grain. Standards Australia International, Sydney.

[3] AS/NZS 1605.1:2006: Methods for sampling and analysing timber preservatives and preservative-treated timber, Part 1: General requirements, sampling, and determination of sapwood and heartwood presence. Standards Australia International, Sydney.

[4] AS/NZS 1605.2:2006: Methods for sampling and analysing timber preservatives and preservative-treated timber, Part 2: Determination of preservative penetration by spot tests. Standards Australia International, Sydney.

[5] AS/NZS 2098.2:2012: Methods of test for veneer and plywood. Method 2: Bond quality of plywood (chisel test). Standards Australia International, Sydney.

[6] AS/NZS 2098.6: 2006 Methods of test for veneer and plywood. Method 6: Depth of peeler checks in veneer and plywood. Standards Australia International, Sydney.

[7] AS/NZS 2098.7:2006 Methods of test for veneer and plywood. Method 7: Density of veneer and plywood. Standards Australia International, Sydney.

[8] AS/NZS 4266.21:2004: Reconstituted wood-based panels—Methods of test—Over-cure or under-cure of resin. Standards Australia International, Sydney.

[9] ASTM D 3359 - 02: Standard test methods for measuring adhesion by tape test. ASTM International, West Conshohocken, PA.

[10] ASTM D 3363 – 00: Standard test method for film hardness by pencil test. ASTM International, West Conshohocken, PA.

[11] ASTM D 4138 - 94 (2001): Standard test methods for measurement of dry film thickness of protective coating systems by destructive means. ASTM International, West Conshohocken, PA.

[12] ASTM D 4541 - 02: Standard test method for pull-off strength of coatings using portable adhesion testers. ASTM International, West Conshohocken, PA.

[13] ASTM D 5266 – 99 (2005): Standard practice for estimating the percentage of wood failure in adhesive bonded joints. ASTM International, West Conshohocken, PA.

[14] ASTM D 4442 – 07: Standard test methods for direct moisture content measurement of wood and wood-base materials. ASTM International, West Conshohocken, PA.

[15] ASTM E 562 – 08e1 (2008): *Standard test Method for Determining Volume Fraction by Systematic Manual Point Count*. ASTM International, West Conshohocken, PA.

[16] Baker, A. J. 1974: Degradation of wood by products of metal corrosion. Forest Service Research Paper FPL 229, Forest Products Laboratory, U.S. Department of Agriculture. Madison, WI.

[17] Bergmann, R. 2010: Drying and Control of moisture content and dimensional changes. Chapter 13: *Wood Handbook—Wood as an Engineering Material.* General Technical Report FPL-GTR-190. Department of Agriculture, Forest Service, Forest Products Laboratory. (www.fpl.fs.fed.us/documnts/fplgtr/fpl_gtr190.pdf).

[18] Blackledge, R. D. (Editor). 2007: *Forensic Analysis on the Cutting Edge.* John Wiley & Sons, Hoboken, NJ.

[19] Böttcher, P. 2004: Oberflächenbehandlung von Holz und Holzwerkstoffen (*Surface finishes and treatments for wood and wood-based products*). Stuttgart: Ulmer Verlag.

[20] BS 6566-8 1985 (withdrawn): Plywood—Specification for bond performance of veneer plywood, Appendix C. Knife test for determining bond quality. British Standards, London.

[21] Bulian, F.; Graystone, J. A. 2009: *Wood Coatings—Theory and Practice.* Elsvier, Amsterdam, Oxford.

[22] Cassens, D. L. 2002: *Quality Control in Lumber Purchasing: Lumber Stress/Casehardening.* Purdue University Forestry and Natural Resources. NEW 4/02, 4 p.

[23] Cassens, D.; Len, Y.; McCabe, G. 2003: Face check development in veneered furniture panels. *Forest Products Journal* 53(10): pp. 79–86.

[24] Christiansen, A. W.; Knaebe, M. 2004. Diagnostic guide for evaluating surface distortions in veneered furniture and cabinetry. General Technical Report FPL-GTR-143. Madison, WI: U.S. Department of Agriculture, Forest Service, Forest Products Laboratory.

[25] Colina, H.; Roux, S. 2000: Experimental model of cracking induced by drying shrinkage. *European Physical Journal* 1(2): pp. 189–194.

[26] De Meier, M.; Militz, H. 2000: Wet adhesion of low-VOC coatings on wood: A quantitative analysis. *Progress in Organic Coatings* 38(3–4): pp. 223–240.

[27] De Meier, M.; Thurich, K.; Militz, H. 1998: Comparative study on penetration characteristics of modern wood coatings. *Wood Science and Technology* 32: pp. 347–365.

[28] Derrick, M. R.; Stulik, D.; Landry, J. M. 1999: *Infrared Spectroscopy in Conservation Science.* The Getty Conservation Institute. ISBN 0-89236-469-6. (http://www.getty.edu/conservation/publications_resources/pdf_publications/infrared_spectroscopy.html).

[29] Dibuz, J. J., Shelton, F. J. 1967. Glueline identification. Forest Products Journal 17(10): pp. 20–22.

[30] EN 13779 2008: Ventilation for non-residential buildings—Performance requirements for ventilation and room-conditioning systems. European Committee for Standardization, Brussels.

[31] EN 1310 1997: Round and sawn timber—Method of measurement of features. European Committee for Standardization, Brussels.

[32] EN 13183-1 2002: Moisture content of a piece of sawn timber—Determination by oven dry method. European Committee for Standardization, Brussels.

[33] Frihart, C. 2009: Adhesive groups and how they relate to the durability of bonded wood. *Journal of Adhesion Science and Technology* 23: pp. 601–617.

[34] Frihart, C. R.; Hunt, C. G. 2010: Adhesives with Wood Materials, Bond Formation and Performance. Chapter 10: *Wood Handbook—Wood as an Engineering Material.* General Technical Report FPL-GTR-190. Department of Agriculture, Forest Service, Forest Products Laboratory. (www.fpl.fs.fed.us/documnts/fplgtr/fpl_gtr190.pdf).

[35] Gereke, T.; Hass, P.; Niemz, P. 2009: Moisture-induced stresses and *distortions* in *spruce* cross-laminates and composite laminates. Holzforschung, 64, pp. 127–133.

[36] Glass, S. V.; Zelinka, S. L. 2010: Moisture relations and physical properties of wood. Chapter 4: *Wood Handbook—Wood as an Engineering Material.* General Technical Report FPL-GTR-190. Department of Agriculture, Forest Service, Forest Products Laboratory. (www.fpl.fs.fed.us/documnts/fplgtr/fpl_gtr190.pdf).

[37] Goisman und Kaplan, 1994: An experimental study of cracking induced by desiccation. *Europhysics Letters* 25(6): pp. 415–420.

[38] Goodell, B. S.; Morrell, J. J. 2013: Predicting the Rate of Decay, and the Potential for Misinterpretation of Proper Scientific Method. International Research Group on Wood Protection. IRG/WP 13-20525.

[39] Hernandez, R. 2009: *Review of 2009 MLB Baseball Bat Regulations.* Technical Note. RockBats, LLC. (http://www.woodbat.org/woodbat.pdf).

[40] Hoadley, R. B. 1990: Identifying Wood: Accurate Results with Simple Tools. *The Taunton Press,* 223 p. ISBN 0-942391-04-7.

[41] Illy, A.; Grüll, G.; Spitaler, I. 2011: Farbänderung durch Holz-Klebstoff-Reaktion (*Color changes caused by wood-adhesive reaction*). Holz-Zentralblatt 33: p. 805 and p. 806.

[42] ISO 12466-1 2007: Plywood—Bonding quality—Test methods. Annex B: Bond testing and evaluation using the chisel/knife testing procedures. International Organization for Standardization, Geneva.

[43] ISO 2811-2 1997: Paints and varnishes—Determination of density—Immersed body (plummet) method. International Organization for Standardization, Geneva.

[44] ISO 15184 2012: Paints and varnishes—Determination of film hardness by pencil test. International Organization for Standardization, Geneva.

[45] ISO 2409 2013: Paints and varnishes—Cross-cut test. International Organization for Standardization, Geneva.

[46] JAS JPIC-EW SE08-02 MAFF 936 2008: Japanese Agricultural Standard for Flooring. Japanese Plywood Inspection Corporation, Tokyo.

[47] Katzengruber, R.; Jeitler, G.; Schickhofer, G. 2005: Interim report on Project P03_qm online of holzbau forschungs GmbH: Zerstörungsfreie Qualitätssicherung von keilgezinktem Konstruktionsvollholz mittels einer Zugprüfung (*Non-destructive quality assurance of finger-jointed structural timber based on a tension proof loading procedure*). Presentation at NDT Congress, Hanover.

[48] Kirch, W.; Engwicht, A. 2005: Fehldiagnosen und Patientensicherheit (*Misdiagnoses and patient safety*). Springer Verlag.

[49] Kirker, G.; Glaeser, J. 2011: Salt Damage to Wood. Techlines – 06-2011-001.

[50] Klyosov, A. A. 2007: *Wood-Plastic Composites.* John Wiley & Sons, Hoboken, NJ.

[51] Koch, G. 2008: *Discolouration of Wood in the Living Tree and During Processing.* In: Proceedings of the COST E53 Workshop "End user's needs for wood material and products", Delft, The Netherlands, pp. 11–18.

[52] Koch, G. 1999: Sekundäre Veränderungen im Holz dynamisch beanspruchter Fichten (Picea abies [l.] Karst.) aus immisionsbelasteten und windbelasteten Hochlagenbeständen (*Secondary changes in wood of dynamically stressed spruce (Picea abies [l.] Karst.) of high-altitude stands due to air pollution and wind exposure*). German Federal Research Centre for Forestry and Forest Products reports. Kommissionsverlag Buchhandlung Max Wiedebusch, Hamburg.

[53] Kretschmann, D. E. 2010: Mechanical Properties of Wood. Chapter 5: *Wood Handbook—Wood as an Engineering Material*. General Technical Report FPL-GTR-190. Department of Agriculture, Forest Service, Forest Products Laboratory. (www.fpl.fs.fed.us/documnts/fplgtr/fpl_gtr190.pdf).

[54] Larson, Philip R.; Kretschmann, David E.; Clark, Alexander III; Isebrands, J. G. 2001: *Formation and Properties of Juvenile Wood in Southern Pines: A Synopsis*. General Technical Report FPL-GTR-129. Madison, WI: U.S. Department of Agriculture, Forest Service, Forest Products Laboratory, 42 p. (http://www.fpl.fs.fed.us/documnts/fplgtr/fplgtr129.pdf).

[55] Leavengood, S.; Funck, J. W.; Reeb, J. E. 2011: A note on face veneer checking in maple plywood. *International Wood Products Journal* 2(2): pp. 120–123.

[56] Lukowsky, D. 2013: Coming undone—Do climate extremes really cause delamination? *Hardwood Floors Magazine*, December 2013/January 2014, pp. 37–39. (http://www.hardwoodfloorsmag.com/FTFDJ14/).

[57] Lukowsky, D.; Buddruhs, N. 2013: Inhibitory extractives or nutrients? Mould growth on coated heartwood and sapwood of Pinus sylvestris and Gleditsia triacanthos. International Research Group on Wood Protection. IRG/WP 13-10806.

[58] Lukowsky, D.; Hora, G. 2002: Pretreatments of wood to enhance the performance of outdoor coatings. Adler, H. -J.; Potje-Kamloth, P. (Eds.). *Macromolecular Symposia*, Wiley-VCH, Weinheim, pp. 77–85.

[59] Mattheck, C. 2011: Thinking Tools after Nature. ISBN-13: 978-3923704750.

[60] Meinlschmidt, P. 2005: Thermographic detection of defects in wood and wood-based materials. 14th International Symposium of Nondestructive Testing of Wood, Hanover, Germany (May 2nd–4th 2005). (http://www.ndt.net/article/v11n01/meinlschmidt/meinlschmidt.pdf).

[61] Niemz, P. 2005: Brinellhärte ausgewählter Holzarten (*Brinell hardness of selected wood species*). Holz-Zentralblatt 61: p. 799.

[62] NWFA 2011: Problems, causes and cures of hardwood floors. National Wood Flooring Association. Technical Publication No. C 200.

[63] Odegaard, N.; Carroll, S.; Zimmt, W. 2000: *Material Characterization Tests for Objects of Art and Archeology*. Archetype Publications, London.

[64] ÖNORM B 3803 2006: Protection of timber used in buildings—Coatings of dimensionally stable wooden outdoor building components—Minimum requirements and testing. Austrian Standards Institute, Vienna.

[65] Payne, J. A., McCormick, A. V., Francis, L. F. 1997: In situ stress measurement apparatus for liquid applied coatings. *Review of Scientific Instruments* 68(12): pp. 4564–4568.

[66] Pillow, M. Y. 1949: *Studies of Compression Failures and their Detection in Ladder Rails*. Report No. 1733, Forest Service, Forest Products Laboratory Madison, WI.

[67] Plath, L. 1965: Zusammenhänge zwischen Verleimung und Lackierung bei der Möbelfertigung (*Interrelations between adhesive bonding and coating in furniture production*). Adhäsion 9: pp. 57–61.

[68] Plath, L. 1967: Mikroskopische Untersuchungen an Polyester-Lackierungen auf Holzoberflächen (*Microscopic examination of polyester coatings on wood surfaces*). Holz als Roh- und Werkstoff 25: pp. 25–31.

[69] Plath, L. 1969: Konstruieren mit Holzwerkstoffen. Die Verleimung von Sperrholz und Holzspanplatten (*Designing with wood-based products. Adhesive bonding of plywood and particleboard*). *Holz-Zentralblatt* 95: pp. 2359, 2362.

[70] prEN 12169 (2006): Criteria for the assessment of conformity of a lot of sawn timber. European Committee for Standardization, Brussels.

[71] Rapp, A.O.; Behrmann, K. 1998: Preparation of wood for microscopic analysis after decay testing. *Holz als Roh- und Werkstoff* 56: pp. 277–278.

[72] Rapp, A.O.; Sudhoff, B.; Pittich, D. 2011: Schäden an Holzfußböden (*Failures in wood flooring*). Schadensfreies Bauen, 29. Fraunhofer IRB Verlag.

[73] Saferstein, R. 2004: *Criminalistics: An Introduction to Forensic Science*. Pearson Education, Inc. Upper Saddle River, NJ.

[74] Sandermann, W.; Lüthgens, M. 1953: Untersuchungen über Verfärbungen von Hölzern (*Investigation into discoloration of woods*). Holz als Roh- und Werkstoff 11: pp. 435–440.

[75] Sandermann, W.; Dietrichs, F. H.; Puth, M. 1960: Über die Trocknungsinhibierung von Lackanstrichen auf Handelshölzern (*On the Inhibition of Drying with Finished Timbers*). Holz als Roh- und Werkstoff 18: pp. 63–75.

[76] SANS 6121 2004: Bond quality of laminated timber products (glulam)—Chisel test. Standards South Africa, Pretoria.

[77] Schmidt, O. 2010: *Wood and Tree Fungi: Biology, Damage, Protection, and Use*, 334 p. ISBN 978-3-642-06875-1. Springer-Verlag.

[78] Schriever, E. 1981: Einfacher qualitativer Nachweis von Harnstofformaldehydharzen und Diisocyanaten in Holzwerkstoffen (*Simple qualitative detection of urea formaldehyde resins and diisocynates in wood-based panels*). Holz als Roh- und Werkstoff, 39(6): pp. 227–229.

[79] Weldon, Dwight G. 2009: *Failure Analysis of Paints and Coatings*. John Wiley & Sons, Hoboken, NJ.

[80] Welling, J. (Ed.) 2010: Dried Timber—how to specify correctly. (www.timberdry.net).

[81] Wiemann, M. C.; Bergman, R. D.; Knaebe, M.; Bowe, S. A. 2009. *Exploring Methods for Prevention of Oxidative Stain in Soft Maple*. Res. Pap. FPL–RP–654. Madison, WI: U.S. Department of Agriculture, Forest Service, Forest Products Laboratory. 6 p.

[82] Willför, S.; Hemming, J.; Reunanen, M.; Eckerman, C.; Holmbom, B. 2003: *Lignans and Lipophilic Extractives in Norway Spruce Knots and Stemwood*. Holzforschung 57(1): pp. 27–36.

[83] Willför, S.; Hemming, J.; Reunanen, M.; Holmbom, B. 2003: *Phenolic and Lipophilic Extractives in Scots Pine Knots and Stemwood*. Holzforschung, 57(4): pp. 359–372.

[84] Williams, R. S. 2010: Finishing of wood. Chapter 16: *Wood Handbook—Wood as an Engineering Material.* General Technical Report FPL-GTR-190. Department of Agriculture, Forest Service, Forest Products Laboratory. (www.fpl.fs.fed.us/documnts/fplgtr/fpl_gtr190.pdf).

[85] Živković, V.; Arnold, M.; Radmanović, M.; Richter, K.; Turkulin, H. 2014: Spectral sensitivity in the photodegradation of fir wood (Abies alba Mill.) surfaces: Colour changes in natural weathering. *Wood Science Technology*, 48, pp. 239–252.

[86] Zosars, A. Kennedy, M. J. 1994: Safer alternative reagents for colour differentiation of heartwood and sapwood. The International Research Group on Wood Protection, Stockholm. IRG/WP 94-20028.

Index

Printed by Libri Plureos GmbH in Hamburg,
Germany

9 780071 839372